Guidance Manual for Disposal of Chlorinated Water

The mission of the Awwa Research Foundation is to advance the science of water to improve the quality of life. Funded primarily through annual subscription payments from over 1,000 utilities, consulting firms, and manufacturers in North America and abroad, AwwaRF sponsors research on all aspects of drinking water, including supply and resources, treatment, monitoring and analysis, distribution, management, and health effects.

From its headquarters in Denver, Colorado, the AwwaRF staff directs and supports the efforts of over 500 volunteers, who are the heart of the research program. These volunteers, serving on various boards and committees, use their expertise to select and monitor research studies to benefit the entire drinking water community.

Research findings are disseminated through a number of technology transfer activities, including research reports, conferences, videotape summaries, and periodicals.

Guidance Manual for Disposal of Chlorinated Water

Prepared by:
Maria Tikkanen, Ph.D.
John H. Schroeter, P.E.
East Bay Municipal Utilities District
375 Eleventh Street
Oakland, CA 94607

Lawrence Y.C. Leong, Ph.D.
Rajagopalan Ganesh, Ph.D.
Advanced Technologies Group
Kennedy/Jenks Consultants, Inc.
2151 Michelson Drive, Ste 100
Irvine, CA 92612

Sponsored
Awwa Research Foundation
6666 West Quincy Avenue
Denver, CO 80235

Published by the Awwa Research Foundation and
American Water Works Association

Disclaimer

This study was funded by the Awwa Research Foundation (AwwaRF). AwwaRF assumes no responsibility for the content of this research study reported in this publication or for the opinions or statements of fact expressed in the report. The mention of the trade names for commercial products does not represent or imply the approval or endorsement of AwwaRF. This report is presented solely for informational purposes.

This study was completed with the assistance of East Bay Municipal Utility District (EBMUD) and nine other water utilities, and is intended to be used as a guidance document for use by other water utilities. It is not a substitute for, or a legal interpretation of Federal, State or local regulations. Users of this manual are cautioned to refer directly to applicable rules and regulations and to contact governing agencies to obtain additional guidance and clarification. In addition, users of this manual must test any devices or procedures contained herein for ability to meet applicable regulatory/permit requirements, as results may vary depending on the discharge and site-specific conditions. Accordingly, users of this manual do so at their own risk and are solely and exclusively responsible for any consequences resulting from such use.

Library of Congress Cataloging-in-Publication Data

Guidance manual for disposal of chlorinated water/prepared by Maria Tikkanen ... [et al.].
  p. cm.
  Includes bibliographical references.
  ISBN 1-58321-143-8
1. Water treatment plants – Waste disposal. 2. Sewage disposal plants – Waste disposal.
3. Water – Purification – Chlorination – Environmental aspects. 4. Chlorine –Toxicology.
I. Tikkanes, Maria.

TD899.W3 G85 2001
628.1'62'0286–dc21   2001046410

Copyright © 2001
by
Awwa Research Foundation
and
American Water Works Association
Printed in the U.S.A.

Printed on recycled paper

ISBN 1-58321-143-8

CONTENTS

LIST OF TABLES	xi
LIST OF FIGURES	xiii
FOREWORD	xv
ACKNOWLEDGMENTS	xvii
EXECUTIVE SUMMARY	xix
CHAPTER 1 INTRODUCTION	1
Chlorination of Potable Waters	1
Release of Chlorinated Water and Concerns	2
Dechlorination of Water Releases	3
Objective	4
CHAPTER 2 REVIEW OF DECHLORINATION REGULATIONS	5
United States Federal Regulations	5
Technology Based Effluent Limits	5
Water Quality Standards Based Discharge Limits	6
State Regulations	8
Regulations for Residual Chlorine Concentrations	8
State Permitting Programs	14
Associated Regulations Impacting Dechlorination	14
Canadian Federal Regulations	27
Canada Water Quality Guidelines	28
Provincial Regulations	29
Alberta	29
British Columbia	29
Manitoba	30
New Brunswick	31
Newfoundland	31
Nova Scotia	32
Ontario	32
Prince Edward Island	32

Saskatchewan	33
Quebec	33
Summary of Canadian Regulations	33
Summary	34

CHAPTER 3 TYPES OF CHLORINATED WATER RELEASES ... 37

Introduction	37
Planned Releases	39
Unplanned Releases	39
Emergency Releases	40
Summary	43

CHAPTER 4 CURRENT DISPOSAL PRACTICES ... 47

Introduction	47
Non-chemical Methods for Chlorine Dissipation	48
Retention in Holding Tanks	48
Land Application of Chlorinated Water	49
Discharge of Chlorinated Water for Groundwater Recharge	50
Discharge Through Hay Bales and Other Natural Obstructions	50
Discharge to Storm Sewers	50
Discharge of Chlorinated Waters in Sanitary Sewers	51
Dechlorination Using Activated Carbon	52
Dechlorination Using Chemicals	53
Sulfur Dioxide (SO_2)	53
Sodium Thiosulfate ($Na_2S_2O_3$)	55
Sodium Sulfite (Na_2SO_3)	58
Sodium Bisulfite ($NaHSO_3$)	60
Sodium Metabisulfite ($Na_2S_2O_5$)	63
Calcium Thiosulfate (CaS_2O_3)	64
Dechlorination Chemical Summary	66
Chemical Feed Techniques	71
Gravity Feed Method	71

Chemical Metering Pumps	72
Venturi Injector Systems	72
Spray Feed Systems	74
Flow-through Systems	74
Automatic Tablet Dispensers	75
Other Flow-through Systems	76
Flow Control Measures	76
Berms	76
Swale	76
Ditches	77
Redirection Pipe	77
Field Methods for Residual Chlorine Measurement	77
Water Quality Test Strips	78
Swimming Pool Test Kits	79
Orthotolidine Indicator Kit	79
Field Colorimetric Test Kits	80
DPD Titration Method	81
Amperometric Titration Method	81
Analytical Interference	83
Summary	84
Dechlorination Information in AWWA Standards	86
Current Dechlorination Practices by Utilities	87
U.S. Utility Practicing Extensive Dechlorination	87
U.S. Utilities Practicing Limited Dechlorination	93
Dechlorination Practices of Utilities not Participating in This Program	94
Dechlorination of Cooling Water Releases	96
Canadian Utility Practicing Extensive Dechlorination	98
Canadian Utility Practicing Limited Dechlorination	102
Summary of Utility Dechlorination Practices	103

CHAPTER 5 NEW TECHNOLOGIES ... 111

Introduction ... 111

Dechlorination Chemicals ... 111

    Ascorbic Acid (Vitamin C) ... 111

    Sodium Ascorbate ... 113

    Hydrogen Peroxide ... 114

New Treatment Media ... 115

    Catalytic Carbon ... 115

    Redox Alloy Media for Dechlorination ... 117

Dechlorination Agent Feed Techniques ... 119

    Dechlor Mat/Strip for Dechlorination Using Tablets ... 119

    Dechlor Diffuser ... 119

    Tablet Diffuser Dechlorinator ... 123

    Sodium Sulfite Tablet/Bag Suspension Dechlorination ... 124

    Tablet/Plastic Pipe Dechlorination ... 127

    Tablet/T –Cage Dechlorination ... 128

    Portable Dechlorinator ... 132

    Chlorination/Dechlorination Unit by Water and Wastewater Technologies, Inc., WA ... 134

    Dechlorination Feeder Unit by City of Salem, OR ... 137

Chlorine Monitoring Method ... 138

    Water Quality Strip Using Tetramethylbenzidine Indicator ... 138

CHAPTER 6 FIELD DECHLORINATION STUDIES ... 141

Introduction ... 141

Dechlorination Studies at Tacoma Waters ... 142

    Background ... 142

    Objectives ... 144

    Work Plan ... 144

    Sample Collection and Analysis ... 148

    Results ... 149

    Field Dechlorination Studies at Bureau of Water Works, Portland 157
        Background 157
        Objectives 157
        Work Plan 158
        Results 159
    Comparison of Free and Combined Chlorine Neutralization 166
    Dechlorination Field Studies at EBMUD 167
        Experimental Conditions 167
        Test Series Descriptions 168
        Test Series Findings 170
        Summary of EBMUD Study 181
    Summary of Tacoma, Portland and EBMUD Field Test Results 182

CHAPTER 7 RECOMMENDATIONS FOR FUTURE WORK 189
    Introduction 189
        Passive Chlorine Removal 189
        Dechlorination Efficiencies of Various Chemicals/Media Under
            Identical Conditions 190
        Effectiveness of Each Chemical Under Different Conditions 191
        Efficiency of Various Chemical Feeders 191
        Improvement in Analytical Procedures 192
        Development of Industry Standard/BMP 192

APPENDIX A UTILITIES/COMPANIES THAT PROVIDED INFORMATION
    FOR THIS REPORT 193
REFERENCES 197
ABBREVIATIONS 201

TABLES

2.1	State regulatory information pertaining to effluent chlorine concentrations	9
2.2	State permit programs for chlorinated water disposal	16
2.3	pH-dependent values of the CMC (acute criterion)	22
2.4	Temperature and pH-dependent values of the CCC (chronic criterion) for fish early life stages (ELS) present	23
2.5	Temperature and pH-dependent values of the CCC (chronic criterion) for fish early life stages (ELS) absent	24
2.6	Approved and working criteria for water quality, British Columbia Ministry of Environment, Lands and Parks (1995)	31
2.7	Provincial regulatory information pertaining to effluent chlorine water	34
3.1	Summary of AWWA disinfection requirements	38
3.2	Types of potable water discharges containing chlorinated water	40
3.3	Categories of chlorinated water releases	43
4.1	Parts of dechlorination chemical required to neutralize one part of free chlorine in distilled water	68
4.2	Regulatory information for various dechlorination chemicals	69
4.3	Comparison of various agents for dechlorination of potable water releases	70
4.4	Comparison of chlorine measurement techniques	85
4.5	Amounts of chemicals required to neutralize various residual chlorine concentrations in 100,000 gal (378.5 M^3) of water	86
4.6	Summary of utility dechlorination practices	106
5.1	Dechlorination procedures for releases from trenches during main breaks	122
5.2	Procedure for dechlor diffuser use	123
5.3	Feasibility of WSSC feed techniques for waters released from different sources	131
5.4	Equipment required for construction of portable dechlorinator	132
6.1	Brief summary of field test conditions at the three sites	143
6.2	Typical water quality characteristics of finished water at Tacoma Water	145
6.3	Chemical feed rate for Tacoma Waters during the field dechlorination studies	147

6.4	pH of water samples during dechlorination at Tacoma Waters	155
6.5	Estimated amounts of dechlorination chemicals remaining in the dechlorination water at Tacoma	156
6.6	Typical water characteristics at Portland Water Bureau	159
6.7	Estimated amounts of dechlorination chemicals remaining in the dechlorinated water at Portland	164
6.8	Ammonia concentrations during dechlorination at Portland	165
6.9	EBMUD dechlorination field testing schedule	168
6.10	Dechlorination using 16 and 20 sodium sulfite tablets in a mesh bag at EBMUD	173
6.11	Dechlorination of EBMUD water using sodium sulfite tablets in tablet dispensers	177
6.12	Dechlorination using ascorbic acid powder at EBMUD	180
6.13	Dechlorination using sodium thiosulfate crystals at EBMUD	180
6.14	Summary of dechlorination trends observed in the field tests	184
A.1	Addresses of agencies whose methodologies are presented in this report	194
A.2	Addresses of companies that provided information for this report	195

FIGURES

4.1	'Bazooka' Venturi dechlorination feeder by Arden Industries (patent pending)....	73
5.1	Dechlor strip for dechlorination of trenches during main breaks......................	120
5.2	Dechlor mat for dechlorination of trenches during main breaks.........................	121
5.3	Diffuser for dechlorination of hydrant or blowoff waters....................................	121
5.4	WSSC tablet/diffuser dechlorinator	125
5.5	WSSC tablet/bag suspension dechlorinator	126
5.6	WSSC plastic pipe dechlorinator	129
5.7	WSSC Tablet/T-cage dechlorinator	130
5.8	Dechlorination chemical feed unit by City of Blue Springs, Missouri	133
5.9	Dechlorination chemical feed unit by Water and Wastewater Technologies, Inc. Bellingham, Washington	135
5.10	Dechlorinator feeder unit by City of Salem, OR...	137
6.1	Field dechlorination test arrangement at Tacoma Water facility	146
6.2	Schematic of field test arrangements and sampling points at Tacoma Water Facility	149
6.3	Chlorine concentrations at Tacoma City Water when stoichiometric concentrations of dechlorination chemicals were used to neutralize chlorine in potable water from a hydrant	150
6.4	Chlorine concentrations at Tacoma Waters when twice the stoichiometric concentrations of dechlorination chemicals were used to neutralize chlorine in potable water from a hydrant	150
6.5	Dissolved oxygen concentrations in Tacoma City waters when stoichiometric concentrations of dechlorination chemicals were used to neutralize chlorine in potable water from a hydrant	152

6.6	Dissolved oxygen concentrations in Tacoma City waters when twice the stoichiometric concentrations of dechlorination chemicals were used to neutralize chlorine in potable water from a hydrant	153
6.7	Field dechlorination test arrangement at Portland Water Bureau facility	161
6.8	Schematic dechlorination test arrangement and sampling points at Portland Water Bureau facility	162
6.9	Chlorine concentrations of Portland Water Bureau water when stoichiometric concentrations of dechlorination chemicals were used to neutralize chlorine in potable water from a hydrant	162
6.10	pH of Portland Water Bureau water when stoichiometric concentrations of dechlorination chemicals were used to neutralize chlorine in potable water from a hydrant	163
6.11	Chlorine and pH levels in EBMUD water when 1 sodium sulfite tablet was placed across the flow (100 gpm)	171
6.12	Chlorine and pH levels in EBMUD waters when 12 sodium sulfite tablets were placed across the flow (100 gpm)	172
6.13	Chlorine, DO and pH levels in EBMUD waters when 28 sodium sulfite tablets were placed across the flow (50 gpm)	172
6.14	Chlorine, DO and pH levels at various points along the flow path using 1 sodium sulfite tablet. The flow rate of the EBMUD water was 100 gpm	175
6.15	Chlorine, DO and pH levels at various points along the flow using 4 sodium sulfite tablets. The flow rate of EBMUD water was 100 gpm	176
6.16	Chlorine and pH levels when 20 sodium sulfite tablets were placed in tablet dispensers across 475 gpm flow of EBMUD water	178
6.17	Chlorine, DO and pH levels when 1 sodium sulfite tablet (Norweco) was placed across 100 gpm flow of EBMUD water	179

FOREWORD

The Awwa Research Foundation is a nonprofit corporation that is dedicated to the implementation of a research effort to help utilities respond to regulatory requirements and traditional high-priority concerns of the industry. The research agenda is developed through a process of consultation with subscribers and drinking water working professionals. Under the umbrella of a Strategic Research Plan, the Research Advisory Council prioritizes the suggested projects based upon current and future needs, applicability, and past work; the recommendations are forwarded to the Board of Trustees for final selection. The foundation also sponsors research projects through the unsolicited proposal process; the Collaborative Research, Research Applications, and Tailored Collaboration programs; and various joint research efforts with organizations such as the U.S. Environmental Protection Agency, the U.S. Bureau of Reclamation, and the Association of California Water Agencies.

This publication is a result of one of these sponsored studies, and it is hoped that its findings will be applied in communities throughout the world. The following report serves not only as a means of communicating the results of the water industry's centralized research program, but also as a tool to enlist the further support of the nonmember utilities and individuals.

Projects are managed closely from their inception to the final report by the foundation's staff and large cadre of volunteers who willingly contribute their time and expertise. The foundation serves a planning and management function and awards contracts to other institutions such as water utilities, universities, and engineering firms. The funding for this research effort comes primarily from the Subscription Program, through which water utilities subscribe to the research program and make an annual payment proportionate to the volume of water they deliver and consultants and manufacturers subscribe based on their annual billings. The program offers a cost-effective and fair method for funding research in the public interest.

A broad spectrum of water supply issues is addressed by the foundation's research agenda: resources, treatment and operations, distribution and storage, water quality and analysis, toxicology, economics, and management. The ultimate purpose of the coordinated effort is to

A broad spectrum of water supply issues is addressed by the foundation's research agenda: resources, treatment and operations, distribution and storage, water quality and analysis, toxicology, economics, and management. The ultimate purpose of the coordinated effort is to assist water suppliers to provide the highest possible quality of water economically and reliably. The true benefits are realized when the results are implemented at the utility level. The foundation's trustees are pleased to offer this publication as a contribution toward that end.

Edmund G. Archuleta. P.E.

Chair, Board of Trustees

Awwa Research Foundation

James F. Manwaring, P.E.

Executive Director

Awwa Research Foundation

ACKNOWLEDGMENTS

The authors of this report are indebted to the following water utilities and individuals for their cooperation and participation in this project:

City of Portland Bureau of Water Works, Portland, OR, Curt Ireland
Tacoma Public Utilities, Tacoma, WA, Khis Suravallop
Department of Public Utilities, City of Naperville, Naperville, IL, Allan Poole
The City of Mesa, Mesa, AZ, William McCarthy
Public Services Board, El Paso Water Utilities, El Paso, TX, Douglas Rittmann
Cincinnati Water Works, Cincinnati, OH, David Hartman
Office of Environmental Services, Broward County, Pompano Beach, FL, Jerry Baker
Regional Municipality of Ottawa-Carlton, Ottawa, Ontario, Canada, Ian Douglas
South Central Connecticut Regional Water Authority, New Haven, CT, Paul Dufour
City of Salem, Salem, OR, Sophia Hobet

The advice of Project Advisory Committee (PAC) – including Jodye Levy Russell, Washington Suburban Sanitary Commission, Laurel, Maryland; Doug McQuarrie, Greater Vancouver Regional District, Burnaby, British Columbia, Canada; Kanval Oberoi, Charleston Commission of Public Works, South Carolina and Daniel Guillory, Metropolitan Water District of Southern California, Los Angeles, California– and the help of initial AWWARF project manager, Chris Rayburn, are appreciated.

EXECUTIVE SUMMARY

BACKGROUND

Dechlorination is practiced by certain U.S. and Canadian water utilities when discharging chlorinated water from new construction, distribution system repairs, water main breaks, hydrant testing and other water utility operations. The volume of water released and the concentration of residual chlorine vary significantly with the source and nature of the discharge. Due to changes in the regulatory approach that include a more ecological and watershed perspective, these practices have been subjected to an increased level of scrutiny. The regulatory review can become more significant as some of the secondary impacts of the treated discharge are more completely understood. For example, over-application of dechlorination chemicals may deplete oxygen concentrations in the receiving streams. In addition, a rise in pH due to dechlorination may increase the concentration of toxic non-ionized ammonia in the receiving streams.

Most written material that is available to water utilities concerning current dechlorination practices is summarized in utility reports. The reports are not readily available to other utilities that have to deal with dechlorination issues. There is currently no industry guidance or standardization similar to that for disinfecting mains and reservoirs (e.g., AWWA Standard C651-92, Disinfecting Water Mains).

The Greater Vancouver Regional District (GVRD) has created two best management practices (BMPs) and a final report titled 'Chlorine Monitoring and Dechlorination Techniques Handbook'. They have posted this information on their website (http://www.gvrd.bc.ca/services/water/chlorlin/index.html#guidelines) and this information can be downloaded and incorporated by other utilities. Other utilities that may have standard operating procedures to address the disposal of chlorinated waters have not made this information so widely available.

This guidance manual summarizes available information on the disposal of chlorinated water, in a centralized, easily available manner. This should save time and effort when a utility has to develop a program to deal with these issues.

OBJECTIVE AND APPROACH

A primary objective of this project is to evaluate current practices and to prepare a guidance manual for disposal of chlorinated water from potable water sources by utilities in the United States (U.S.) and Canada. The major tasks include the following:

- compilation of regulations that restrict the release of chlorinated water into the environment in the United States and Canada
- identification of the activities releasing chlorinated waters
- documentation of dechlorination techniques currently followed
- review of new dechlorination technologies, and
- summary of field dechlorination studies

This manual has been prepared through the active participation of ten utilities in geographically diverse locations located throughout the U.S. and Canada. In addition, information obtained from three Project Advisory Committee (PAC) member utilities has also been included. These utilities differ in their use of disinfection agents (eight of the utilities use free chlorine and five use combined chlorine) and source waters (seven utilities primarily use surface water, one uses only groundwater and the rest use a combination of ground and surface waters). In addition, the population served by these utilities also varies significantly (from 110,000 to 1,800,000). Some utilities wholesale most of their waters and the rest of the utilities retail most of their waters. All these differences have an impact on the water management practices and hence, in the quantity and character of the chlorinated water released.

Field dechlorination studies were conducted at Tacoma City Water, WA; Portland Water Bureau, OR and East Bay Municipal Utility District (EBMUD), CA to evaluate dechlorination efficiencies and water quality impacts of various chemicals. The results of these studies, and their implications are included in this report.

COMPILATION OF DECHLORINATION REGULATIONS

U.S. Dechlorination Regulations

The United States Environmental Protection Agency (USEPA) has established water quality criteria (1986) for total residual chlorine (TRC) concentration for receiving streams based on acute and chronic toxic effects for aquatic life. Under the acute toxicity criterion, the maximum chlorine concentration for the protection of aquatic life in receiving fresh waters is 19 µg/L (microgram per liter). To meet the acute toxicity criterion, the 1-hour average chlorine concentration should not exceed 19 µg/L more than once every three years on the average. Under the chronic toxicity criterion, the maximum allowable chlorine concentration in the fresh water is 11 µg/L. The four-day average concentration should not exceed 11 µg/L more than once every three years on the average.

In addition, many state regulatory agencies have developed numerical or toxicity based chlorine limits for water discharged into receiving streams. Many states require utilities to obtain permits for discharging specific types of chlorinated waters into receiving streams. A review of permit programs of various states indicated some differences in the approach to regulate potable water releases. In California, Oregon, Nevada, Maryland and West Virginia, the residual chlorine concentration in all waters discharging into the receiving streams should not exceed 0.1 mg/L (milligram per liter). Colorado, Connecticut, Tennessee, Kentucky, Wisconsin and Wyoming have more than one general permit to regulate various chlorinated water releases. However, these permits do not include all potable water discharges. In many midwestern states including Iowa, Kansas, Illinois and Michigan and some Southeastern states including Arkansas, Georgia and Louisiana, no general or individual permit program is in place for potable water releases. However, utilities in these states are required to meet the water quality criteria of receiving streams while discharging chlorinated waters.

Canadian Dechlorination Regulations

The task force on water quality for the Canadian Environmental Quality Guidelines (1987) has proposed a water quality criterion of 2 µg/L of total residual chlorine for receiving streams. All provincial regulatory agencies have adopted this chlorine concentration as the water quality criterion for receiving streams. Programs to regulate chlorine concentrations in waters discharged into receiving streams differ in various provinces. British Columbia and Ontario require all water releases to contain less than 0.002 µg/L total residual chlorine. Newfoundland has a chlorine discharge limit of 1 mg/L. Other provinces either determine the chlorine concentration of discharge on a case-by-case basis or require discharges to comply with water quality criteria for the receiving streams.

TYPES OF CHLORINATED WATER RELEASES

Potential types of chlorinated water releases during various utility operations have been compiled to develop a treatment/disposal approach for various release situations. The releases can be classified as planned, unplanned and emergency releases. Planned releases result from operation and maintenance activities such as disinfection of mains, testing of hydrants and routine flushing of distribution system lines and mains for maintenance. Unplanned releases occur from activities such as main breaks, leaks and overflows. Activities such as water main flushing in response to higher than allowable coliform counts and taste and odor complaints from the public are examples of emergency releases of chlorinated waters. Although planned releases sometimes may contain higher concentrations of chlorine, these discharges are easier to control and hence, easier to dechlorinate. Unplanned and emergency releases are harder to dechlorinate due to limitations in response time, staff availability and the difficulty in containing these waters.

Based on the residual chlorine concentration, the sources can be classified as low-level disinfection residual (< 4 mg/L) and high-level disinfection residual (super chlorinated, > 4 mg/L) waters. While most waters released from distribution systems contain lower concentrations of residual chlorine, disinfection of new or repaired mains may generate waters containing significantly higher concentrations of residual chlorine. The concentration of residual chlorine may alter the dechlorination approach.

Furthermore, the volume and frequency of releases can also impact the selected method of dechlorination. Hence, relevant information for each type of chlorinated water is also compiled in this report. Finally, the compiled types of releases are grouped into the following categories to facilitate development of guidance for dechlorination:

- planned, low chlorine, high flow releases
- planned, low chlorine, moderate flow releases
- planned, low chlorine, low flow releases
- planned, high chlorine, moderate flow releases
- unplanned, low chlorine, moderate flow releases
- unplanned, low chlorine, high flow releases and
- unplanned, low chlorine and low flow releases.

CURRENT DECHLORINATION PRACTICES

Various non-chemical and chemical methods for dechlorination, chlorine measurement techniques and dechlorination practices of participating utilities are summarized below.

Passive, Non-Chemical Dechlorination Methods

Some of the passive, non-chemical techniques for the safe disposal of chlorinated waters include retention in holding ponds for chlorine dissipation, land application of chlorinated water, discharge of chlorinated water for groundwater recharge, discharge through hay bales and other natural obstructions and discharge into sanitary sewers. The primary advantage of dissipating chlorine passively is that such a process does not involve chemical addition. As a result, utilities do not have to be concerned with effects of neutralizing chemicals in the receiving streams. In addition, the cost, health and safety concerns related to storage, transportation and handling of these chemicals can be avoided.

However, passive dechlorination methods have several limitations. First, chlorine dissipation through natural decay is very slow. Second, some activities produce a large volume of chlorinated waters requiring a larger holding tank for dissipation by retention. Also, an appropriate site must be available close to the point of chlorinated water release for disposal by land application and groundwater recharge methods. The primary limitations in disposal through sanitary sewers include availability of a sanitary sewer near the point of chlorinated water release and the available capacities of the sanitary sewer and the wastewater treatment plant.

Dechlorination Using Chemicals

Chemicals such as sulfur dioxide, sodium bisulfite, sodium metabisulfite, sodium sulfite, sodium thiosulfate and calcium thiosulfate are used for neutralizing residual chlorine. While most of the chemicals can readily neutralize chlorine from potable waters, some chemicals can cause potential health concerns if not handled properly. For example, sulfur dioxide is a hazardous gas and is not well suited for field dechlorination applications. Sodium bisulfite and sodium metabisulfite are skin, eye or respiratory tract irritants. They are harmful if swallowed or inhaled. In addition, sulfite-based dechlorination chemicals can cause water quality concerns by depleting dissolved oxygen concentrations. Also, some dechlorination chemicals produce hydrochloric acid and decrease water pH. While most of the chemicals are available in powder or crystal forms, sodium sulfite is the only dechlorination chemical currently available in tablet form.

Chlorine Measurement Techniques

Various techniques are available for free and combined chlorine measurement. Most field methods for measuring chlorine concentrations use an indicator that alters the color of the water sample, which is then compared to a set of colored standards. Swimming pool test kits, orthotolidine indicator kits and field pocket colorimetric test kits are among the test kits that rely upon this technique. Test kits can measure free, combined and total chlorine by adding N,N-diethyl-p-phenylenediamine (DPD) or orthotolidine indicator to the water.

Current Dechlorination Practices of Participating Utilities

Dechlorination practices of participating utilities, this project's Project Advisory Committee (PAC) member utilities, some other utilities and industrial facilities have been summarized. EBMUD, Washington Suburban Sanitation Commission (WSSC), GVRD and Portland Water Bureau dechlorinate all planned water releases. In addition, these utilities have active program to address unplanned releases. Tacoma Water, the Regional Municipality of Ottawa-Carlton and certain other utilities dechlorinate selected chlorinated water releases. Utilities such as City of El Paso, TX; Broward County, FL; City of Mesa, AZ do not dechlorinate any of their water releases.

A common method of disposal of chlorinated water for many utilities is discharge into sanitary sewers. If this is not a viable option, utilities use various chemical agents for dechlorination. Many utilities in California, the GVRD, the Region of Ottawa-Carlton and an industrial facility in Kentucky use sodium thiosulfate for dechlorination. Portland Water Bureau and EBMUD use sodium thiosulfate for selected applications. EBMUD and WSSC use sodium sulfite tablets for most of their dechlorination needs. Almost all of the water utilities use Hach pocket colorimeters for monitoring residual chlorine concentrations.

NEW TECHNOLOGIES

Chemicals

Ascorbic acid and sodium ascorbate have been recently identified as chemicals that remove chlorine effectively. Initial studies indicate that these chemicals do not deplete dissolved oxygen (DO) concentrations in the water significantly. However, ascorbic acid has been reported to decrease the pH of water by 2 to 3 units in some cases. To minimize this problem, some utilities are using sodium ascorbate for dechlorination. Sodium ascorbate is more expensive than most other dechlorination chemicals, but has less adverse impact on water quality. While ascorbic acid and sodium ascorbate powders have a shelf life of over a year, once in solution they decompose within a few days.

Devices

WSSC has developed various devices/methods for dechlorination using sodium sulfite tablets. Included are "Tablet/Diffuser" for dechlorination of hydrant discharge and blowoff water, "Tablet/Bag Suspension" for blowoff water in deep manholes, "Tablet/Plastic pipe dechlorinator" for end-wall type blowoffs and "Tablet/Cage" for hydrant flushing and other high velocity discharges. The Tablet/Cage device can handle a flow rate of up to 2000 gpm. EBMUD uses "Dechlor Strips" and "Dechlor Mats" for dechlorination of discharges from trenches during main breaks and other such flows. These Dechlor strips and Mats facilitate effective contact between the flow and the tablets during dechlorination. For releases from hydrants and blowoffs, a diffuser chamber has been developed that optimizes water/tablet contact during dechlorination. Water and Wastewater, Inc. and City of Salem, (OR) have developed venturi-based devices for delivering dechlorination chemical solutions. The devices can handle a flow rate of up to 1000 gpm (gallons per minute).

BEST MANAGEMENT PRACTICES

One of the stated objectives of the research is to develop BMPs for various dechlorination activities. However, during the preparation of this report, it was observed that sufficient information is not currently available to develop comprehensive BMPs. In particular, the following concerns remain:

- Not all the dechlorination chemicals have been tested under identical situations to evaluate dechlorination efficiencies and water quality impacts.

- Many chemicals have not been tested for various chlorinated water releases, feed rates, forms, etc.

- Very little information is available on the disposal of superchlorinated water.

- Very limited information is available on dechlorination of groundwaters.

- Complete information on the toxicity of dechlorination chemicals is not available.

- Dechlorination chemical feed methods have not been optimized for various conditions.

Hence, BMPs for disposal of chlorinated waters have not been developed in this report. However, field studies were conducted to obtain preliminary information on dechlorination of free and combined chlorine from hydrant water and concurrent water quality impacts. In addition, suggestions for further works required to develop BMPs have been made in this report.

FIELD DECHLORINATION STUDIES

Field tests were conducted at Tacoma Water, WA and Portland Water Bureau, OR to evaluate dechlorination using six different chemicals in solution. Tacoma Water uses free chlorine, and Portland Water Bureau uses combined chlorine for disinfection. In these studies, a 1% solution of each of the chemicals was made and delivered at feed rates just sufficient to neutralize all the chlorine in the water. In addition, tests were conducted using dechlorination chemicals at twice the amount required for chlorine neutralization. Upon chemical addition, chlorine residual, pH and DO were measured along the flow. Field studies conducted at EBMUD in 1998 used sodium sulfite tablets, ascorbic acid powder and sodium thiosulfate crystals to dechlorinate hydrant discharges. EBMUD uses combined chlorine for disinfection. At all three test sites hydrant waters originated from surface water sources rather than from groundwaters.

The results from all three tests have been summarized. The results indicated the following:

- All of the chemicals tested in solution, tablet or powder form were able to neutralize free and combined chlorine to below 0.1 mg/L.

- Stoichiometric concentrations of dechlorination chemicals in solution removed more than 90 % of residual chlorine.
- The DO concentration decreased by 1 mg/L when the stoichiometric amount of sodium metabisulfite was added.
- A decrease in DO concentrations (~ 1.0 mg/L) was observed when twice the stoichiometric amounts of sodium metabisulfite, sodium sulfite or sodium thiosulfite were used.
- Dechlorination reactions of calcium thiosulfate were slower than those of other chemicals tested.
- Reactions of sodium/calcium thiosulfate with chloramine were slower than those with free chlorine.
- While ascorbic acid and sodium thiosulfate solutions neutralized chlorine effectively, when used in powder/crystal form, they dissolved rapidly, causing water quality concerns by adversely impacting pH and oxidation reduction potential (ORP) levels.
- Sodium sulfite tablets effectively dechlorinated chloramines from water released from a hydrant. When water from a hydrant was released at 300 gpm, one sodium sulfite tablet, contained in a mesh, in the path of the flow was sufficient to dechlorinate 2.0 mg/L of chloramines to below 0.1 mg/L for 45 minutes.
- When a large number of sodium sulfite tablets (28) were placed across a water released from a hydrant at a low floe rate (50 gpm), the DO concentration of the water decreased by 5 mg/L (from 8 mg/L to 3 mg/L), within 25 minutes.
- When 12 tablets were placed across a flow of 100 gpm, the residual chlorine concentration was below 0.1 mg/L for more than 60 minutes. However, when the flow rate was increased to 475 gpm, the residual chlorine concentration increased to values above 0.1 mg/L within 25 minutes. This indicated that, under similar conditions, an increase in flow rate may decrease the length of time the residual chlorine concentrations were below the detection limit.

- While residual chlorine concentrations of less than 0.1 mg/L were measured in several tests, the instrument used and field conditions selected did not yield zero chlorine residual in most cases.

RECOMMENDATIONS FOR FUTURE WORK

The current state of dechlorination practices indicates that more work needs to be performed to better understand the mechanisms and issues involved. First, effectiveness of passive, non-chemical dechlorination, caused by chlorine demand of impurities on various surfaces has not been well documented. Although several chemicals have been recently identified for chlorine neutralization, these chemicals have not been evaluated under various chlorinated water releases. In order to develop an industry standard and provide a Best Management Practices Manual, the following issues must be addressed:

- Passive chlorine removal by chlorine demand exerted by various surfaces must be evaluated.

- Dechlorination efficiencies, economy and water quality impacts of each of the chemicals under identical conditions must be evaluated.

- Effectiveness of the chemicals and equipments must be tested under various release conditions.

- Dechlorination and water quality impacts using various chemicals for superchlorinated water must be evaluated.

- Dechlorination and water quality impacts must be evaluated for groundwaters.

- Ease, economy, adaptability and effectiveness of various tablet and dechlorination chemical feeders must be studied.

CHAPTER 1

INTRODUCTION

CHLORINATION OF POTABLE WATERS

Chlorine has been used as a disinfectant in potable water systems for over 100 years. Although various disinfectants such as ozone, UV light and chlorine dioxide have been used by the water industry, chlorine is the disinfectant of choice for most utilities due to its effectiveness, efficiency, economy of operation and convenience (Connell 1996).

Free chlorine and combined chlorine (chloramines) are the two forms of chlorine widely used for the disinfection of potable waters. Free chlorine is added as chlorine gas or sodium/calcium hypochlorite to the water. The reaction of chlorine in water produces hypochlorous acid and hydrochloric acid.

$$Cl_2 + H_2O \leftrightarrow HOCl + HCl$$

Hypochlorous acid, in turn, dissociates into hydrogen ion and hypochlorite ion.

$$HOCl \leftrightarrow H^+ + OCl^-$$

Hypochlorous acid is the stronger oxidizing agent (disinfectant) of the two forms. The degree of dissociation of hypochlorous acid is dependent on the pH of the water. At pH 7.0 and room temperatures, nearly 70 % of the chlorine remains as hypochlorous acid. A slight increase in pH significantly increases the hypochlorite ion concentration. At a pH of 8.0, approximately 70 % of the chlorine is converted to hypochlorite ion.

Free chlorine is highly reactive and *relatively* unstable. Utilities using free chlorine for disinfection often use secondary chlorination stations to maintain residual chlorine concentrations in the potable water as regulated by the Clean Water Act (CWA).

Combined chlorine is produced by adding chlorine and ammonia to water. Usually a chlorine-to-ammonia ratio of 3:1 to 6:1 is used. Ammonia forms monochloramine, dichloramine and trichloramine species with chlorine, as per the following reactions:

$$HOCl + NH_3 \leftrightarrow NH_2Cl + H_2O$$
$$HOCl + NH_2Cl \leftrightarrow NHCl_2 + H_2O$$
$$HOCL + NHCl_2 \leftrightarrow NCl_3 + H_2O$$

Monochloramine is a more effective oxidizing agent than di- and trichloramines. The distribution of chloramine species is dependent on the pH, temperature, chlorine to ammonia ratio and contact time. In potable water systems, monochloramine is often the predominant species present. Since chlorine is bound by ammonia and hydrogen, combined chlorines are less aggressive, more persistent and react more slowly with oxidizable materials and bacteria. (Connell 1996). Under certain conditions, free chlorine may react with organic substances in water to form carcinogenic trihalomethanes (THMs). The potential for formation of THMs using combined chlorines is significantly lower than that using free chlorine. Odor problems using combined chlorine are reported to be less than that of free chlorine. For these reasons some utilities prefer combined chlorine for disinfection.

RELEASE OF CHLORINATED WATER AND CONCERNS

Chlorinated water from potable water sources is released to the environment through activities such as water main flushing, disinfection of new mains and storage tanks, hydrant testing, main breaks, filter backwash, etc. Although chlorine protects humans from pathogens in the water, it is highly toxic to various forms of aquatic life. Free and combined chlorines are reported to be toxic to some aquatic species at or below 0.1 mg/L. The Clean Water Act has defined the water quality criterion for total residual chlorine to be as low as 11 µg/L for some waters. In addition to chlorine, ammonia is also toxic to several aquatic species at very low concentrations (0.1 - 0.2 mg/L) (USEPA 1984). The toxicity of ammonia is dependent on its speciation in the water. Unionized ammonia (NH_3) is reported to be more toxic than the ionized form (NH_4^+). An increase in pH usually increases the concentration of unionized ammonia in the water. Federal and State regulations have developed water quality criteria for unionized ammonia and total ammonia to be as low as 0.002 and 0.08 mg/L, respectively, for some waters.

DECHLORINATION OF WATER RELEASES

Wastewater treatment plants dechlorinate treated effluent prior to release into receiving streams. For a long time sulfur dioxide gas had been the chemical of choice for dechlorination. Sodium bisulfite and sodium metabisulfite solutions were also used. However, sulfur dioxide is rated as a hazardous gas and must be handled with extreme care. Sodium bisulfite and metabisulfite are corrosive and bisulfite has a tendency to crystallize below room temperature. Hence, utilities dechlorinating potable waters often use less hazardous chemicals such as sodium sulfite and sodium thiosulfate in tablet or powder forms. Recently, chemicals such as ascorbic acid, sodium ascorbate and calcium thiosulfate have been found to neutralize chlorine effectively. Some utilities are currently evaluating dechlorination of potable waters using these chemicals.

More importantly, many water utilities often use passive, non-chemical methods such as discharge to sanitary sewers for disposal of chlorinated waters. Impurities such as organics, iron and sulfide in the sanitary sewer exert a chlorine demand and neutralize chlorine in the released water.

Dechlorination is an evolving practice in the water industry. Currently, there is no industry guidance or AWWA standard for disposal of chlorinated water. The AWWA standard for disinfection of water mains (C651-92) provides some information on disposal of chlorinated waters. This information is not a part of the standard. Regulations and permit programs for disposal of chlorinated waters vary widely among the states and provinces.

Although several chemicals are used by the utilities to neutralize chlorine in potable waters, impacts of these chemicals upon the water quality of receiving streams are not well documented. At elevated concentrations, sodium sulfite and sodium thiosulfate can deplete dissolved oxygen concentrations in receiving streams. Ascorbic acid, at high doses, may decrease the water pH below permissible levels. Furthermore, disposal of chlorinated water to sanitary sewers requires consideration of sewer capacity and chlorine impact on the operation of the wastewater treatment plant. In addition, discharge fees are often imposed that can be significant. Discharge without dechlorination to storm sewers is often unacceptable since the discharge goes directly into a receiving stream.

This project attempts to summarize the current dechlorination practices by various utilities in a centralized, easily available manner. This will save time and effort when a utility has to develop a program to deal with dechlorination issues. In addition, the manual provides preliminary information to help communicate and shape policy with the regulators. This report identifies chlorine neutralizing techniques and constraints from an operational perspective when dealing with chlorinated water releases. Furthermore, preliminary field studies were conducted to enhance our understanding of dechlorination techniques and chemicals and the results are included in this report.

OBJECTIVE

The primary objective of this project is to prepare a guidance manual for disposal of chlorinated water from potable water sources. The major tasks, as described in the project scope of work with the American Water Works Association Research Foundation (AWWARF), include the following:

- Compilation of regulations that restrict the release of chlorinated water into the environment in the United States and Canada

- Identification of activities releasing chlorinated waters

- Documentation of dechlorination techniques currently followed

- Review of new dechlorination technologies

- Development of Best Management Practices (BMPs) for various dechlorination activities.[1]

This manual has been prepared through the active participation of ten utilities in geographically diverse locations throughout the United States and Canada.

---

[1] See Chapter 6 for details of this task.

CHAPTER 2

REVIEW OF DECHLORINATION REGULATIONS

This chapter summarizes regulations related to the disposal of chlorine from potable water sources. In addition, regulations for water quality parameters such as ammonia and dissolved oxygen concentrations that are affected by dechlorination are also presented. National Pollution Discharge Elimination System (NPDES) permit programs currently in place for a number of states and provinces are also included. However, regulations or permit programs for the union territories are not included. Regulations for transportation or spills of chlorine (hazardous waste management) and disinfection by-product formation are beyond the scope of this study. Furthermore, regulations related to chlorine gas release are not addressed.

UNITED STATES FEDERAL REGULATIONS

Most of the U.S. federal laws and regulations pertaining to the disposal of chlorinated water are promulgated under the Clean Water Act (CWA) (U.S. Code Title 33, Chapter 26, Water Pollution Prevention and Control) and Code of Federal Regulations, Title 40, Protection of Environment (40 CFR). Under these rules and regulations, the United States Environmental Protection Agency (USEPA) is authorized to maintain and improve the purity and quality of the nation's receiving waters.

In general, the discharge limits for chlorine and other pollutants are determined from two perspectives. The first approach involves regulating pollutant concentrations in effluents based on available technologies. The second approach for effluent limits involves the Water Quality Standards (WQS) for receiving waters. Both of these approaches are described in the following sections.

Technology Based Effluent Limits

40 CFR 122 (Subsections 41 - 44) and 40 CFR 23.5 provide information related to technology based effluent limits. The authority and guidance for the regulations are provided by

Sections 301, 306 and 307 of the Clean Water Act (Sections 1311, 1314 and 1317 of U.S. Code, Chapter 33).

Technology-based treatment represents the minimum level of control that must be imposed in a permit issued under Section 402 of the Clean Water Act (40 CFR 123.5). For dischargers other than publicly owned treatment works (POTWs), the discharge limits are based on the following criteria:

- the Best Practicable Control Technology currently available (BPT)
- the Best Conventional Pollutant Control Technology (BCT) for conventional pollutants
- the Best Available Technology economically achievable (BAT) for all toxic pollutants, or
- BAT for all pollutants, which are neither toxic nor conventional pollutants.

Under the technology-based criteria, the effluent limits are determined by the amounts and characteristics of the pollutants and the degree of effluent reduction attainable through the application of the best practicable control technology. Factors such as the total cost of the application in relation to the effluent reduction benefits are also considered.

Water Quality Standards Based Discharge Limits

Sections 302 and 303 of the CWA (U.S. Code 1312 and 1313, Chapter 33) and 40 CFR 122.44 describe WQS for receiving waters and water quality based effluent standards. The regulations require the USEPA to develop WQS to protect the physical, chemical and biological characteristics of the Nation's waters. When the effluent limitations based on technology are not sufficient to protect the aquatic life of the receiving waters, the limitations based on water quality criteria must be used to define the pollutant concentration in discharge waters.

The WQS set the maximum concentration of pollutants permissible in the receiving waters to protect aquatic life and water quality. Under the authority provided by the CWA, the USEPA has established Water Quality Criteria (1986) for total residual chlorine concentrations that are based on acute and chronic toxicity effects for aquatic life. Under the acute toxicity criterion, the maximum chlorine concentration for the protection of aquatic life in receiving fresh waters is 19 µg/L. To meet the acute toxicity criterion, the 1-hour average chlorine concentration should not exceed 19 µg/L more than once every three years on the average. Under the chronic toxicity criterion, the maximum allowable chlorine concentration in fresh waters is 11 µg/L. The 4-day average concentrations should not exceed this concentration more than once every three years on the average. Total residual chlorine concentration for marine waters are 13 and 7.5 µg/L under acute and chronic toxicity criteria, respectively. Although these chlorine concentrations are below the detection limits of the current analytical methods, they are used in calculating the allowable discharge chlorine concentrations to protect aquatic species.

During the release of chlorinated waters, to determine compliance of WQS, samples for residual chlorine measurement are collected from outside a zone called the "Mixing Zone" in the receiving stream. The mixing zone is that portion of a water body adjacent to an effluent outfall where mixing results in the dilution of the effluent with the water. Water quality criteria may be exceeded in a mixing zone subject to some conditions.

For example, the State of Washington Administrative Code WAC 173-201A-100 indicates that no mixing zone shall be granted unless the supporting information clearly indicates that the mixing zone:

- would not have a reasonable potential to cause a loss of sensitive or important habitat, and
- would not substantially interfere with the uses of the water body, result in damage to the ecosystem or adversely affect public health.

The maximum size of mixing zones in rivers and streams shall comply with the most restrictive combination of the following:

- not extend in a downstream direction for a distance from the discharge port greater than 300 feet plus the depth of water discharge or extend upstream for a distance of 100 feet;

- not utilize greater than 25 % of the flow; and
- not occupy greater than 25 % of the width of the water body.

STATE REGULATIONS

This section summarizes the state regulations related to the discharge of chlorinated water within the U.S. Although not included in the scope of work of this project, permit programs under State Pollution Discharge Elimination Systems have been identified for 36 states and summarized in this section. The regional differences in dechlorination regulations and practices in the United States are also summarized. Table 2.1 summarizes the regulatory agencies responsible for water quality control in each state and the regulatory limits for chlorine residuals. While the regulations of some states define the numerical limit for residual chlorine concentration in the water being discharged, in most cases, the regulations set the water quality criterion for residual chlorine for the receiving streams. The chlorine limits in the discharge water is decided in the permit program in these states. Table 2.1 provides the discharge chlorine limit or the water quality criterion of the receiving streams as provided in the state regulations.

Regulations for Residual Chlorine Concentrations

In this subsection, effluent quality and water quality based state regulations for chlorinated water discharge are summarized. The types of chlorinated water exempted from such regulations are also presented.

Water Quality Standards Based Chlorine Regulations

As shown in Table 2.1, most of the states follow USEPA acute (19 µg/L) and chronic (11 µg/L) toxicity concentrations for chlorine as the regulatory receiving WQS. States such as Indiana and Iowa have adopted a modified acute and chronic WQS. States such as Arizona and New Mexico have no primacy over water quality. They follow Federal regulatory requirements for the discharge of chlorinated water.

Table 2.1

State regulatory information pertaining to effluent chlorine concentrations

State	Regulatory agency	Residual chlorine regulation
Alabama*	Water Division Dept. of Environmental Management	No numeric value. Case by case toxicity based limitation.
Alaska*	Div. of Facility Construction & Operation	Refers to USEPA standard. Toxicity based criteria proposed.
Arizona[†]	Water Quality Division, Dept. of Environmental Quality	Acute[‡] - 11 µg/L residual chlorine Chronic[§] - 5 µg/L residual chlorine
Arkansas*	Department of Pollution Control and Ecology	No numeric value. Case by case toxicity based limitation.
California*	State Water Resources Control Board & Nine Regional Water Quality Control Boards	Discharge limit varies from 0.00 to 0.10 mg/L and includes toxicity-based criteria.
Colorado[†]	Water Quality Control Division, Dept. of Health and Environment	Acute[‡] - 0.019 mg/L residual chlorine Chronic[§] - 0.011 mg/L residual chlorine
Connecticut[†]	Department of Environmental Protection	Refers to USEPA standards.
Delaware[†]	Delaware Dept. of Natural Resources and Env. Control	Chronic[§] 0.011 mg/L residual chlorine
Florida[†]	Division of Water Facilities, Dept. of Environmental Protection	Water Quality Standard 0.01 mg/L
Georgia[†]	Dept. of Natural Resources	Refers to USEPA standard.
Hawaii[†]	Environmental Management Division	Acute[‡] - 0.019 mg/L residual chlorine Chronic[§] - 0.011 mg/L residual chlorine
Idaho[†]	Idaho Division of Environmental Quality	Chronic[§] 0.011 mg/L residual chlorine

(continued)

Table 2.1 (Continued)

State	Regulatory agency	Residual chlorine regulation
Indiana[†,*]	Indiana Department of Environmental Management	Acute[‡] - 19 µg/L residual chlorine Chronic[§] - 11 µg/L residual chlorine Intermittent flow discharge limit 0.2 mg/L
Iowa[†]	Water Quality Bureau, Dept. of Natural Resources	Acute - 35 µg/L residual chlorine Chronic - 10 µg/L residual chlorine
Illinois[†]	Bureau of Water, Environmental Protection Agency	Acute[‡] - 19 µg/L residual chlorine Chronic[§] - 11 µg/L residual chlorine
Kansas[†]	Bureau of Water, Dept. of Health and Environment	Acute[‡] - 19 µg/L residual chlorine Chronic[§] - 11 µg/L residual chlorine
Kentucky[†]	Division of Water	Acute[‡] - 19 µg/L residual chlorine Chronic[§] - 10 µg/L residual chlorine
Louisiana[*]	Department of Environmental Quality	Case by case toxicity based limitation.
Maine[†]	Department of Environmental Protection	Acute[‡] - 19 µg/L residual chlorine Chronic[§] - 11 µg/L residual chlorine
Maryland[*]	Maryland Department of the Environment	< 0.1 mg/L residual chlorine
Massachusetts[†]	Department of Environmental Protection	Acute[‡] - 19 µg/L residual chlorine Chronic[§] - 11 µg/L residual chlorine
Michigan[*]	Department of Environmental Quality	No specific standards for chlorine. Refers to general toxicity test.
Minnesota[†]	Minnesota Pollution Control Agency	Maximum 0.019 mg/L residual chlorine Chronic 0.006 mg/L residual chlorine
Mississippi[†]	Surface Water Division, Dep. of Environmental Quality	Acute[‡] - 19 µg/L residual chlorine Chronic[§] - 11 µg/L residual chlorine

(continued)

Table 2.1 (Continued)

State	Regulatory agency	Residual chlorine regulation
Missouri[†,*]	Department of Natural Resources	Acute[‡] - 19 µg/L - warm water Chronic[§] - 10 µg/L - warm water Chronic - 2 µg/L - cold water Exempts releases with 1 mile travel & 50 fold dilution in losing streams
Montana[*]	Department of Environmental Quality	Acute[‡] - 19 µg/L - warm water Chronic[§] - 10 µg/L - warm water Disinfection and main flush waters are exempted from regulations by statute.
Nebraska[†]	Department of Environmental Quality	Warm water - Acute[‡] - 19 µg/L Chronic[§] - 11 µg/L Cold water - Acute[‡] - 35 µg/L Chronic[§] - 21 µg/L
Nevada[†]	Division of Environmental Protection	Acute[‡] - 19 µg/L residual chlorine Chronic[§] - 11 µg/L residual chlorine
New Hampshire[†]	Department of Environmental Services	Acute[‡] - 19 µg/L residual chlorine Chronic[§] - 11 µg/L residual chlorine
New Jersey[†]	Department of Environmental Protection	Acute[‡] - 19 µg/L residual chlorine Chronic[§] - 11 µg/L residual chlorine
New Mexico[†]	Surface Water Quality Bureau, Dept. of Environment	Acute[‡] - 19 µg/L residual chlorine Chronic[§] - 11 µg/L residual chlorine
New York[†]	Dept. of Environmental Conservation	5 µg/L residual chlorine
North Carolina[†]	Division of Env. Management, Dept. of Health & Natural Resources	17 µg/L residual chlorine
North Dakota[†]	Environmental Health Section, Department of Health	Acute[‡] - 19 µg/L residual chlorine Chronic[§] - 11 µg/L residual chlorine

(continued)

Table 2.1 (Continued)

State	Regulatory agency	Residual chlorine regulation
Ohio[†,*]	Division of Surface Water, Environmental Protection Agency	Acute[‡] - 19 µg/L residual chlorine Chronic[§] - 11 µg/L; or 0.038 mg/L in effluent as per BAT[**]
Oklahoma*	Department of Environmental Quality	No numeric criteria. Refers to acute and chronic toxicity criteria.
Oregon[†]	Water Quality Division, Dept. of Environmental Quality	Acute[‡] - 19 µg/L residual chlorine Chronic[§] - 11 µg/L residual chlorine 0.1 mg/L in effluent[††]
Pennsylvania*	Department of Env. Protection	BAT or 0.5 mg/L in discharge
Rhode Island[†]	Department of Environmental Management	Acute[‡] - 19 µg/L residual chlorine Chronic[§] - 11 µg/L residual chlorine
South Carolina[†]	Department of Environment and Natural Resources	Refers to USEPA Standards
South Dakota[†]	Department of Environment and Natural Resources	Acute[‡] - 19 µg/L residual chlorine Chronic[§] - 11 µg/L residual chlorine
Tennessee[†]	Department of the Environment and Conservation	No numerical value. Refers to USEPA criteria.
Texas[†]	Office of Water Resources, Nat. Res. Conservation Commission	Acute[‡] - 19 µg/L residual chlorine Chronic[§] - 11 µg/L residual chlorine
Utah[†]	Division of Water Quality, Dept. of Environmental Quality	7 day average below 0.011 mg/L
Vermont[†]	Department of Environmental Conservation	Acute[‡] - 19 µg/L residual chlorine Chronic[§] - 11 µg/L residual chlorine
Virginia[†]	Department of Environmental Quality	Acute[‡] - 19 µg/L residual chlorine Chronic[§] - 11 µg/L residual chlorine
Washington[†]	Department of Ecology	Acute[‡] - 19 µg/L residual chlorine Chronic[§] - 11 µg/L residual chlorine
West Virginia*	Division of Environmental Protection	0 residual chlorine in trout waters. 10 µg/L in hydrotesting waters

(continued)

Table 2.1 (Continued)

State	Regulatory agency	Residual chlorine regulation
Wisconsin[†]	Department of Natural Resources	Acute[‡] 19.03 µg/L residual chlorine Chronic[§] - 7.28 µg/L residual chlorine
Wyoming[†]	Water Quality Division	Acute[‡] - 19 µg/L residual chlorine Chronic[§] - 11 µg/L residual chlorine

Source: Information obtained directly from State regulatory agencies mentioned in columns 1 and 2 or obtained from their internet sites.

*   Regulation defines numeric or descriptive criteria for discharge water.
†   Regulation defines residual chlorine limit for the receiving stream.
‡   Water Quality Standard for acute toxicity criteria.
§   Water Quality Standard for chronic toxicity criteria.
**   Concentration determined by Best Available Technology.
††   Concentration defined in Management Practices.

When chlorine or other pollutant concentrations are regulated through the receiving WQS, issues such as existing water quality, proposed use of the stream, types of species in the stream are considered.

Discharge Water Quality Based Regulations

In lieu of WQS criteria, states such as Alabama, Arkansas, Louisiana, Michigan and Oklahoma regulate chlorine concentrations through total effluent toxicity criteria. Various procedures for evaluating effluent toxicity are also prescribed by each state. In California, the San Francisco Bay and San Diego Regional Water Quality Control Boards (RWQCBs) have prescribed toxicity-based criteria.

Some states have set effluent discharge concentrations for the release of chlorinated waters into the environment. The Oregon Department of Environmental Quality (ODEQ), under its Management Practices for Disposal of Chlorinated Water, allows a maximum chlorine concentration of 0.1 mg/L for all discharges into receiving waters of the state. The Los Angeles and Santa Ana RWQCBs in California allow a maximum total residual chlorine concentration of 0.1 mg/L in all waters discharging to state water bodies. The California Central Coast RWQCB allows 0.002 mg/L of chlorine in discharge waters. The State of Colorado regulations suggest that if the limitation attainable by Best Available Technology is more stringent than that indicated by WQS, then the former should be applied. The effluent chlorine concentration based on BAT in Ohio is 0.038 mg/L. Indiana has a total residual chlorine concentration of 0.2 mg/L for intermittent flows.

Although some of these allowable concentrations are not measurable by current analytical capabilities, the regulatory agencies use these concentrations as the allowable concentration for pollutants. Determination of concentrations below detection limits of chlorine, using currently available analytical methods, is considered to represent compliance with the regulations.

State Permitting Programs

In order to meet the receiving water quality criteria for chlorine and other contaminants, state regulatory boards develop a State Pollution Discharge Elimination System program for discharge of waters and wastewaters into receiving streams. Various general and individual permits are issued by each state to regulate release of chlorinated water from potable water sources. In order to reflect the regional differences in dechlorination practices throughout the nation, the types of permits available, waters regulated by the permits and discharges exempted from permit requirements for 36 states are briefly summarized (Table 2.2).

Associated Regulations Impacting Dechlorination

This section addresses non-chlorine regulations that impact dechlorination practices.

Regulations for Ammonia Concentrations

Dechlorination of chloramine residuals may result in the release of ammonia in select discharges. Hence, regulations pertaining to ammonia concentrations are also reviewed in this report. The non-ionized form of ammonia (NH_3) is more toxic to aquatic species than the ionized form (NH_4^+). The speciation of ammonia is dependant on the pH, temperature and alkalinity of the receiving water body. USEPA has determined the maximum allowable acute concentrations (based on the pH of the water) and chronic concentrations (based on pH and temperature of the water) of ammonia (USEPA, 1999). In general, the one-hour average concentration of total ammonia nitrogen (in mg N/L) should not exceed, more than once every three years on average, the acute criterion for the water body. The thirty-day average concentration of total ammonia nitrogen (in mg N/L) should not exceed, more than once every three years on average, the chronic criterion for the water body. In addition, the highest four-day average within the 30-day period should not exceed 2.5 times the chronic criterion. Furthermore, ammonia concentrations are defined based on the type of aquatic habitat present. The criterion is more stringent in bodies where salmonid fish or fish early life stages are present. Many states have adopted the guidance provided by the USEPA to set water quality standards for ammonia concentrations.

The EPA criteria for ammonia are shown in Tables 2.3, 2.4 and 2.5. The criteria for ammonia vary significantly with the pH and temperature of the receiving streams. Under the worst-case scenario (pH 9.0, 30° C), a one-hour average total ammonia concentration must not exceed 0.885 mg/L as N (1.07 mg/L as NH_3) and the one-month average total ammonia should not exceed 0.179 mg/L as N (0.217 mg/L as NH_3).

Water utilities using combined chlorine often maintain a residual chlorine concentration of less than 1.5 mg/L in potable waters, with a chlorine to ammonia ratio of 4:1 to 5:1 by mass. This will result in a total ammonia concentration of approximately 0.4 to 0.5 mg/L. This is less than the acute criteria for total ammonia concentrations under the worst-case scenario (1.07 mg/L as NH_3). In addition, a five-fold dilution of chloraminated water by the receiving stream will result in compliance with chronic ammonia concentration criteria (0.217 mg/L as NH_3) under the most stringent conditions. Hence, for the majority of dechlorination operations, ammonia released from chloramines may be within the regulatory requirements. However, some states may have regulations more stringent than the EPA criteria for selected waters. Caution must be exercised in releasing chloraminated waters into receiving streams under such conditions.

Table 2.2

State permit programs for chlorinated water disposal

State	Chlorine permit information
1. Arkansas	No permit is required for potable water discharges. Regulates chlorine in wastewater discharge through toxicity tests.
2. California	Discharge chlorine concentrations vary from 0.0 to 0.1 mg/L in various regions. Some basin plans prescribe toxicity-based criteria.
3. Colorado	The state has a Minimal Discharge Permit (COG6-600000) for hydrotesting waters and super chlorinated waters generated by disinfection activities. Colorado also has a general permit for major water treatment plant discharges. The regulatory agency is currently planning to include miscellaneous discharges from distribution systems in one of the existing permits. Under the permit requirements, allowable chlorine concentrations in water release vary based on chlorine concentrations and flow rates of receiving streams. However, residual chlorine should never exceed 0.5 mg/L in the released water. All discharges are exempted from permit requirements if released to sanitary sewer.
4. Connecticut	The state has general permits for filter backwash (DEP-PERD-GP-001), and hydrotesting waters (DEP-PERD-GP-009). Chlorine concentration should not exceed 50 µg/L in hydrotesting water releases. Chlorine limits for backwash waters vary from 9 µg/L to 0.9 mg/L depending on stream dilution capacity. The state plans to write a separate permit, or include in one of the existing permits, discharge limits for super-chlorinated waters.
5. Georgia	The state regulates selected discharges such as swimming pool water releases by permit program. Although releases from water distribution system are not exempted from WQS criteria, they are often unregulated. However, the state expects utilities to abide by WQS requirements.
6. Hawaii	Hawaii has a general permit for hydrotesting waters. The permit does not specify effluent chlorine concentrations. However, acute and chronic chlorine criteria of 19 and 11 µg/L, respectively, from various releases are measured at the end of the pipe, not in receiving waters.

(continued)

Table 2.2 (Continued)

State	Chlorine permit information
7. Idaho	Idaho does not have primacy over NPDES discharges (Non-Delegated State). Historically some permits existed for filter-backwash discharges. However, no limit for chlorine was prescribed. Currently, there is no general or individual permit for release of chlorinated water. However, utilities are required to abide by State acute (0.019 mg/L) and chronic (0.011 mg/L) chlorine criteria for receiving streams.
8. Illinois	No permit in place for chlorinated water releases from potable water sources. The state recommends that utilities dechlorinate these releases. However, non-compliance is not pursued as a violation. Individual permits exist for discharges from large mains (> 36 " diameter). The utilities must abide by WQS criteria.
9. Indiana	The state does not have a general permit for chlorinated water releases, but provides individual permits. Chlorine discharge limit of 0.6 mg/L is prescribed.
10. Iowa	No general permit is in place. Individual permits issued. Discharge chlorine limits are based on WQS calculations. Several discharges often go unregulated. Dechlorination is currently not a high priority issue in the state.
11. Kansas	No general or individual permit program is in place for discharge of chlorinated waters. Utilities are instructed to maintain WQS while releasing such waters into streams.
12. Kentucky	There is a general permit for Backwash Waters. For hydrotesting and main disinfection, distribution system related waters, either a general NPDES permit or a temporary one time approval is given. No discharge to receiving streams is allowed without one of the above two. The state specifies no chlorine limit but requires the utilities to abide by WQS criteria and report chlorine monitoring results to state. Occasionally, the regional regulatory office decides on a specific chlorine limit on a case-by-case basis.
13. Louisiana	A draft general permit for Backwash Water is in review. General permit for hydrotesting is in place. No specific limit for chlorine is prescribed. WQS and material safety data sheet (MSDS) based regulation. Other flows are not regulated by permit.

(continued)

Table 2.2 (Continued)

State	Chlorine permit information
14. Maine	No general permit in place. Permits issued individually. Chlorine concentrations based on WQS or BPT concentration of 1.0 mg/L, whichever is more stringent.
15. Maryland	State general permit no. 95-GP-005 for all chlorinated water releases. Discharge chlorine limit is less than 0.1 mg/L.
16. Michigan	No general permit for potable water releases. Utilities are instructed to adhere to AWWA Standards for disinfection/dechlorination practices.
17. Montana	No general permit in place. However, discharge of chlorinated water into receiving streams are discouraged. Montana Department of Environmental Quality (MDEQ) recommends discharge of these waters to percolation ponds or other non-surface water sinks. Potable water discharge to surface waters is regulated through individual permits. Discharge chlorine limit in such an event is 0.5 mg/L.
18. Nebraska	A general permit is in place for hydrotesting. It recommends not using chlorinated water. However, if chlorinated water is used, the releases must not violate WQS criteria. No general permit for other potable water releases. Agencies are generally required to abide by WQS requirements.
19. Nevada	Issue either a temporary permit (6 months) for infrequent events or an individual permit for large utilities. Chlorine limit varies from 0.1 mg/L in stringent situations to up to 4 mg/L, in remote places and where receiving streams are more than 1,000 feet away. The state follows Oregon's system of calculations. No plans for general permit in the horizon. However, waters containing greater than 4 mg/L chlorine traveling more than 1000 ft to reach receiving streams at volumes less than 500 gallons are exempted from permit requirements.
20. New Jersey	Has a general permit for swimming pool release. However, individual permits are provided for hydrotesting waters. Chlorine produced oxidant concentration must be less than 0.1 mg/L. No permit in place for other releases.
21. New Mexico	State has no primacy. USEPA Region VI controls. Water use designation based discharge limits prevail. Up to 1 mg/L chlorine is allowed in ephemeral streams. Discharge limit can be as low as 19 µg/L for sensitive streams. No general permit in place to regulate all discharges.

(continued)

Table 2.2 (Continued)

State	Chlorine permit information
22. New York	New York does not have a general permit for chlorinated water release. The receiving water quality (WQ) criteria for chlorine is 5 µg/L. Utilities need to obtain individual permit for chlorinated water release. The discharge limit for chlorine is based on the chlorinated and receiving water flows. If the flows are such that a dilution of greater than 70:1 can be achieved in the receiving waters, than a discharge limit of 2 mg/L is applied to chlorinated waters. If the dilution is between 70:1 and 30:1 then a receiving water quality limit of 0.1 mg/L is used in discharge water chlorine limit. If the discharge dilution is less than 30:1 then a discharge limit of 25 µg/L is used.
23. North Carolina	Chlorine not regulated for backwash waters from existing facilities. New or expanding facility chlorine discharge limit is 17 ppb. Filter backwash from raw water filtration tank and main flushing and hydrostatic pressure testing waters are exempted from permit requirements. However, these releases must not violate WQS criteria.
24. North Dakota	Permit No. NDG 070000 permits any temporary dewatering activity if chlorine concentration is less than 0.1 mg/L. No permit in place for other distribution system related discharges. Backwash waters are sent to holding ponds or storm/sanitary sewers. AWWA Standards 651-653 practices are recommended.
25. Oregon	The state requires chlorine concentration to be less than 0.1 mg/L in all discharges to the receiving stream. However, waters containing < 4 mg/L chlorine traveling more than 1000 ft and of volumes less than 500 gallons are exempted from permit requirements.
26. South Carolina	The state has three general discharge permits (Water Treatment Plant Treatment Discharges with i) Maximum, ii) Median and iii) the lowest Residual Chlorine limits). The ten year, seven day average (7Q10) data of the receiving water will be used as the basis for deciding the appropriate permit for a discharge. These permits generally refer to water treatment plant discharges.

(continued)

Table 2.2 (Continued)

State	Chlorine permit information
South Carolina (continued)	However, the facilities obtaining these permits are allowed to discharge the following non-storm water discharges with identical chlorine limits: discharge from fire fighting activities, fire hydrant flushings, potable water sources including waterline flushing, irrigation drainage, lawn watering, routine external building washdown not using detergents or other compounds, pavement washwaters containing no toxic or hazardous materials, air conditioning condensate, springs, uncontaminated groundwater and foundation or footing waters. Hydrant and main flushing waters should have no detergents or chemical compounds and the occasional peak chlorine conc. should not exceed 4.0 mg/L. For maximum TRC permit (SCG641000) the monthly and daily averages should not exceed 0.5 and 1.0 mg/L, respectively. For Median TRC permit (SCG 643000) the monthly and daily averages should not exceed 0.25 and 0.5 mg/L, respectively. For the lowest TRC permit the averages should be less than 0.05 mg/L.
27. South Dakota	No permit issued. Utilities need to follow AWWA standard method guidelines.
28. Tennessee	State has NPDES permits for filter backwash water and hydrotesting waters. Backwash water chlorine concentration must be the lower of 1 mg/L or (0.02 + 0.2 X stream dilution) concentration. Discharge from hydrotesting waters must not violate WQS criteria. No general permit in place for flushing & disinfection related issues.
29. Texas	Hydrotesting waters discharge permit is in place by Rule. Individual permits for water releases. Discharge limits vary from 2 mg/L and above. The state has just started evaluate this issue.
30. Utah	A letter of approval is provided in response to requests from utilities. However, there is no general permit in place for these types of releases. Utilities are required to meet WQS and prevent fish kills. A chlorine limit of 0.11 mg/L is prescribed in a letter of approval.
31. Virginia	Virginia does not have general permit for potable water releases. However, utilities are required to meet WQS requirements.
32. Vermont	No general permit in place. All releases regulated by individual permit or administrative orders. Chlorine concentrations decided on a case-by-case basis.

(continued)

Table 2.2 (Continued)

State	Chlorine permit information
33. Washington	No general permit in place. Chlorine concentrations generally regulated through storm water NPDES permit program. Utilities are required to meet WQS criteria.
34. West Virginia	Has a general permit for hydrotesting of new pipelines. Chlorine must not be present if these waters are discharged to trout waters. Chlorine concentrations must be less than 10 µg/L if discharged to non-trout waters. Other water related discharges are not exempted but unregulated. Utilities are required to ensure WQS while releasing such waters.
35. Wisconsin	There is a general permit for discharge of chlorine containing waters. Permits recommend specific dechlorination alternatives. Chlorine concentrations during discharges must be reported to the state.
36. Wyoming	Wyoming doesn't have a general permit, but has a temporary permit for each type of release. Chlorine concentrations will be decided on a case by case basis, but the concentration should never exceed 1.0 mg/L.

The permit discharge limit for ammonia varies from state to state and with the designated use of the water. In general, the permissible one-hour average concentration of the non-ionized ammonia varies from 9 µg/L (as NH_3) in waters bearing sensitive species at $0°$ C, pH 6.5 to 260 µg/L (as NH_3) at $30°$ C, pH 9.0. The permissible four-day average concentration of non-ionized ammonia (as NH_3) in water bearing sensitive species varies from 0.7 µg/L at $0°$ C, pH 6.5 to 42 µg/L at $30°$ C, pH 9.0. The State of Florida has set a maximum non-ionized ammonia limit of less than 0.02 mg/L for select waters and the State of Colorado has set a limit of 0.03 to 0.1 mg/L ammonia, depending on the classification of receiving water. The Central Coast Basin of California permits a maximum non-ionized ammonia concentration of 0.025 mg/L for all effluents discharging to the water bodies of the region.

Table 2.3

pH-dependent values of the CMC (acute criterion)

pH	CMC (mg N/L) Salmonids present	Salmonids absent
6.5	32.6	48.8
6.6	31.3	46.8
6.7	29.8	44.6
6.8	28.1	42.0
6.9	26.2	39.1
7.0	24.1	36.1
7.1	22.0	32.8
7.2	19.7	29.5
7.3	17.5	26.2
7.4	15.4	23.0
7.5	13.3	19.9
7.6	11.4	17.0
7.7	9.65	14.4
7.8	8.11	12.1
7.9	6.77	10.1
8.0	5.62	8.40
8.1	4.64	6.95
8.2	3.83	5.72
8.3	3.15	4.71
8.4	2.59	3.88
8.5	2.14	3.20
8.6	1.77	2.65
8.7	1.47	2.20
8.8	1.23	1.84
8.9	1.04	1.56
9.0	0.885	1.32

Source: USEPA, 1999, Update of Ambient Water Quality Criteria for Ammonia.

Table 2.4

Temperature and pH-dependent values of the CCC (chronic criterion) for fish early life stages present

pH	CCC for fish early life stages present (mg N/L) Temperature (C)								
	0	14	16	18	20	22	24	26	28	30
6.5	6.67	6.67	6.06	5.33	4.68	4.12	3.62	3.18	2.80	2.46
6.6	6.57	6.57	5.97	5.25	4.61	4.05	3.56	3.13	2.75	2.42
6.7	6.44	6.44	5.86	5.15	4.52	3.98	3.50	3.07	2.70	2.37
6.8	6.29	6.29	5.72	5.03	4.42	3.89	3.42	3.00	2.64	2.32
6.9	6.12	6.12	5.56	4.89	4.30	3.78	3.32	2.92	2.57	2.25
7.0	5.91	5.91	5.37	4.72	4.15	3.65	3.21	2.82	2.48	2.18
7.1	5.67	5.67	5.15	4.53	3.98	3.50	3.08	2.70	2.38	2.09
7.2	5.39	5.39	4.90	4.31	3.78	3.33	2.92	2.57	2.26	1.99
7.3	5.08	5.08	4.61	4.06	3.57	3.13	2.76	2.42	2.13	1.87
7.4	4.73	4.73	4.30	3.78	3.32	2.92	2.57	2.26	1.98	1.74
7.5	4.36	4.36	3.97	3.49	3.06	2.69	2.37	2.08	1.83	1.61
7.6	3.98	3.98	3.61	3.18	2.79	2.45	2.16	1.90	1.67	1.47
7.7	3.58	3.58	3.25	2.86	2.51	2.21	1.94	1.71	1.50	1.32
7.8	3.18	3.18	2.89	2.54	2.23	1.96	1.73	1.52	1.33	1.17
7.9	2.80	2.80	2.54	2.24	1.96	1.73	1.52	1.33	1.17	1.03
8.0	2.43	2.43	2.21	1.94	1.71	1.50	1.32	1.16	1.02	0.897
8.1	2.10	2.10	1.91	1.68	1.47	1.29	1.14	1.00	0.879	0.773
8.2	1.79	1.79	1.63	1.43	1.26	1.11	0.973	0.855	0.752	0.661
8.3	1.52	1.52	1.39	1.22	1.07	0.941	0.827	0.727	0.639	0.562
8.4	1.29	1.29	1.17	1.03	0.906	0.796	0.700	0.615	0.541	0.475
8.5	1.09	1.09	0.990	0.870	0.765	0.672	0.591	0.520	0.457	0.401
8.6	0.920	0.920	0.836	0.735	0.646	0.568	0.499	0.439	0.386	0.339
8.7	0.778	0.778	0.707	0.622	0.547	0.480	0.422	0.371	0.326	0.287
8.8	0.661	0.661	0.601	0.528	0.464	0.408	0.359	0.315	0.277	0.244
8.9	0.565	0.565	0.513	0.451	0.397	0.349	0.306	0.269	0.237	0.208
9.0	0.486	0.486	0.442	0.389	0.342	0.300	0.264	0.232	0.204	0.179

Source: USEPA, 1999, Update of Ambient Water Quality Criteria for Ammonia

Table 2.5

Temperature and pH-dependent values of the CCC (chronic criterion) for fish early life stages (ELS) absent

	CCC for fish early life stages absent (mg N/L)									
pH	Temperature									
	0-7	8	9	10	11	12	13	14	15*	16*
6.5	10.8	10.1	9.51	8.92	8.36	7.84	7.35	6.89	6.46	6.06
6.6	10.7	9.99	9.37	8.79	8.24	7.72	7.24	6.79	6.36	5.97
6.7	10.5	9.81	9.20	8.62	8.08	7.58	7.11	6.66	6.25	5.86
6.8	10.2	9.58	8.98	8.42	7.90	7.40	6.94	6.51	6.10	5.72
6.9	9.93	9.31	8.73	8.19	7.68	7.20	6.75	6.33	5.93	5.56
7.0	9.60	9.00	8.43	7.91	7.41	6.95	6.52	6.11	5.73	5.37
7.1	9.20	8.63	8.09	7.58	7.11	6.67	6.25	5.86	5.49	5.15
7.2	8.75	8.20	7.69	7.21	6.76	6.34	5.94	5.57	5.22	4.90
7.3	8.24	7.73	7.25	6.79	6.37	5.97	5.60	5.25	4.92	4.61
7.4	7.69	7.21	6.76	6.33	5.94	5.57	5.22	4.89	4.59	4.30
7.5	7.09	6.64	6.23	5.84	5.48	5.13	4.81	4.51	4.23	3.97
7.6	6.46	6.05	5.67	5.32	4.99	4.68	4.38	4.11	3.85	3.61
7.7	5.81	5.45	5.11	4.79	4.49	4.21	3.95	3.70	3.47	3.25
7.8	5.17	4.84	4.54	4.26	3.99	3.74	3.51	3.29	3.09	2.89
7.9	4.54	4.26	3.99	3.74	3.51	3.29	3.09	2.89	2.71	2.54
8.0	3.95	3.70	3.47	3.26	3.05	2.86	2.68	2.52	2.36	2.21
8.1	3.41	3.19	2.99	2.81	2.63	2.47	2.31	2.17	2.03	1.91
8.2	2.91	2.73	2.56	2.40	2.25	2.11	1.98	1.85	1.74	1.63
8.3	2.47	2.32	2.18	2.04	1.91	1.79	1.68	1.58	1.48	1.39
8.4	2.09	1.96	1.84	1.73	1.62	1.52	1.42	1.33	1.25	1.17
8.5	1.77	1.66	1.55	1.46	1.37	1.28	1.20	1.13	1.06	0.990
8.6	1.49	1.40	1.31	1.23	1.15	1.08	1.01	0.951	0.892	0.836
8.7	1.26	1.18	1.11	1.04	0.976	0.915	0.858	0.805	0.754	0.707
8.8	1.07	1.01	0.944	0.885	0.829	0.778	0.729	0.684	0.641	0.601
8.9	0.917	0.860	0.806	0.756	0.709	0.664	0.623	0.584	0.548	0.513
9.0	0.790	0.740	0.694	0.651	0.610	0.572	0.536	0.503	0.471	0.442

Source: USEPA, 1999, Update of Ambient Water Quality Criteria for Ammonia

* At 15 C and above, the criterion for fish ELS absent is the same as the criterion for fish ELS present.

Dissolved Oxygen Concentration Regulations

Dechlorination using reducing agents such as sodium bisulfite and sodium metabisulfite may deplete oxygen concentrations in the receiving waters. Hence, regulations related to dissolved oxygen concentrations in streams have also been reviewed. The WQS for minimum dissolved oxygen concentrations appear to vary from state to state and by the type of water use. The standard also varies for cold and warm waters. The cold-water requirements are more stringent than warm water requirements. In most states, receiving waters are classified according to their designated uses. The dissolved oxygen (DO) standard is more stringent for waters designated for drinking water supply and contact recreational activities than those allotted for non-contact uses such as navigation. Minimum dissolved oxygen concentration for select fresh waters in Oregon, identified by the Department of Environmental Quality as providing for salmonid spawning (during the periods from spawning until fry emergence from the gravels), is as high as 11 mg/L. A concentration as low as 2.5 mg/L is permitted for select waters of Arizona.

However, in most cases, the minimum dissolved oxygen concentration for warm water streams is about 5.0 to 6.0 mg/L and the concentrations in cold waters vary from 6.0 to 9.0 mg/L. In Alabama, Idaho, Iowa, Louisiana, Missouri and Tennessee, the minimum DO concentration for most receiving streams is 5.0 mg/L.

pH Regulations

Many dechlorinating agents produce hydrochloric and sulfuric acids while neutralizing chlorine. If sufficient alkalinity is not present in the water, dechlorination may reduce the discharge/receiving water pH. In addition, the efficiency of some dechlorinating agents appears to vary with water pH. Hence, regulations related to pH in receiving waters have also been evaluated. In most states, receiving water pH varies with the use classification of the receiving streams. The acceptable pH levels vary from 4.0 to 9.0. However, most of the state WQS require a receiving water pH between 6.0 and 8.5. Furthermore, many regulatory agencies require that the pH of the receiving stream should not be altered by more than 0.2 to 0.5 units upon release of chlorinated or other waters. Many state permit programs require discharge water pH to be between 5.0 and 9.0.

Regulations for Water Quality Limited Receiving Streams

All state waters, based on designated use and benefits, are categorized into various classes. The water quality requirements for receiving streams are determined based on the category of use. The standards are more stringent for waters used for drinking and contact sports than for those used for navigation and other non-contact, non-consumptive uses. In some cases, the existing quality of water may not meet the quality criteria set based on its use classification. These waters are designated as water quality limited waters. 'Water Quality Limited' bodies can mean the following:

- a receiving stream which does not meet in-stream WQS during the entire year or defined season even after the implementation of standard technology; or
- a receiving stream, which achieves and is expected to continue to achieve in-stream WQS but utilizes higher than standard technology to protect beneficial uses.

'Water Quality Limited' bodies also refers to:

- a stream for which there is insufficient information to determine if WQS are being met; or
- receiving streams that would not be expected to meet WQS during the entire year or a defined season using standard technologies.

Hence, a more stringent limitation on the pollutant discharge concentration than specified by general WQS is defined for water quality limited bodies. Several states have identified the streams that are water quality limited and set more stringent WQS than those defined by USEPA. Water quality of impaired streams is protected by developing 'Total Maximum Daily Loads (TMDLs)' and 'Waste Load Allocations (WLAs)' of pollutant discharges. TMDLs represent the sum of existing and/or projected point source, non-point source and background pollutant loads for a water body. TMDLs set and allocate the maximum amount of a pollutant that may be introduced into water that still ensures attainment and maintenance of WQS.

The following procedure is generally followed in evaluating pollutant concentrations in discharge waters. Initially, the TMDL or Waste Load Allocation (WLA) of pollutants is determined for a receiving body, depending on the conditions of that water body. The WLA is the portion of the receiving water's loading capacity that is allocated to one of its existing or future point sources of pollution. For example, Ohio uses the following formula, along with safety factors, to calculate the WLA (OAR 3745-2-05).

$$WLA = \frac{WQC*(Q_{eff} + Q_{up}) - Q_{up}*WQ_{up}}{Q_{eff}}$$

Where,

    WLA - Waste Load Allocation (µg/L)

    WQC - Water Quality Criterion (µg/L)

    Q_{eff} - Effluent Flow Rate (gal/day)

    Q_{up} - Stream design flow (gal/day)

    WQ_{up} - Background Water Quality (µg/L)

Other states use similar procedures to arrive at effluent pollutant concentrations. The effluent chlorine concentration is then determined based on the effluent flow rate and WLA.

For example, the total nitrogen concentration (including the ammonia nitrogen concentration) in select waters discharging into water quality limited streams in Oregon should not be higher than 0.040 mg/L. Discharge of chlorinated waters to these streams must meet the more stringent water quality objectives.

CANADIAN FEDERAL REGULATIONS

Among the Canadian federal statutes, the Fisheries Act and the Criminal Code of Canada regulates chlorinated potable water releases. Section 36(3) of the Federal Fisheries Act prohibits any person, unless and otherwise authorized by the Act, from depositing deleterious substances of any type in water frequented by fish. Under Section 40(2), every person found to have contravened Section 36(3) is guilty of a violation and may be subject to a fine as high as $1,000,000 or imprisonment for not more than 3 years, or both. Section 40(5) states that "deposit" takes place whether or not any act or omission resulting in the deposit is intentional. Section 42 states that the persons who at any time own the deleterious substance or have the charge, management and control of it, or who otherwise cause or contribute to the deposit, are, subject to exception, jointly liable for all expenses incurred by the government in the measures taken to reverse any adverse effect of the deposit. Section 42(4) further states that the above liability is absolute and does not depend upon proof of fault or negligence. Section 35 prohibits anyone, unless authorized by this Act, from an undertaking that results in harmful alteration of fish habitat. Section 78.3 provides that in any prosecution for an offense under the fisheries act, it will be sufficient for the Crown to establish that the offense was committed by an agent or employee of the person charged, whether or not the employee or agent is identified or has been prosecuted for the offense. The only exception is when it can be established that the employee or agent committed the offence without the knowledge or the consent of the person charged.

The Criminal Code of Canada makes it a criminal offense to deposit offensive volatile substances into any place likely to alarm, inconvenience or discomfort any person or damage any property.

Canada Water Quality Guidelines

The Guidelines and Standards Division of the Canadian Council of Resources and Environment Ministers provides nationally approved, science-based indicators of environmental quality. The primary focus of this group is to develop national guidelines for water quality, sediment quality, soil quality and aquatic tissue residues. These guidelines are recommended numerical or narrative limits for a variety of substances and environmental quality parameters, which, if exceeded, may impair the health of Canadian ecosystems. Guidelines are mandated federally under the Canadian Environmental Protection Act (CEPA) and nationally under various federal-provincial agreements (e.g., Canadian Council of Ministers of the Environment (CCME), Great Lakes Water Quality Agreement).

The task force on Water Quality has developed and promulgated Canadian Environmental Quality Guidelines (1987) numerical concentrations or narrative statements. These guidelines are required under Part 1, Section 8, of the Canadian Environmental Protection Act. The guidelines propose a water quality criterion of 2 µg/L of total residual chlorine for receiving streams. Many provincial regulatory agencies have adopted this chlorine concentration as the WQS criterion for receiving streams. However, the criteria for chlorine and chloramine concentrations are currently being reviewed under the 1999 water quality guidelines.

PROVINCIAL REGULATIONS

As described earlier, provincial regulatory agencies have adopted a water quality criterion of 2 µg/L for the receiving streams. However, the process for regulating chlorinated water release varies from province to province. The regulatory/permit process for each province is briefly described below.

Alberta

The province of Alberta has a Water Quality Objective of 0.002 mg/L for chlorine. Under the provincial Standard Guidance for Maintenance of Water, Wastewater and Storm Drainage Systems chlorinated waters are prohibited from discharge directly into streams and storm sewers. Dechlorination is required prior to release. The province mandates these requirements while providing permission to utilities to operate water treatment and distribution systems. No additional permit is provided during each release event. However, currently some chlorinated water releases go unregulated into the receiving streams.

British Columbia

The Ministry of Environment defines regulations for maintaining environmental quality in British Columbia (BC). A comprehensive report on disposal of chlorinated water in BC has been developed by the Greater Vancouver Regional District (GVRD). This report describes various provincial regulations governing dechlorination of potable water releases. Excerpts from the GVRD report are presented in this section.

B.C. regulations stipulate different discharge chlorine limits for continuous and intermittent flows. For continuous flows, the total residual chlorine (TRC) concentrations in discharge waters should not exceed 0.002 mg/L. For intermittent flows, the regulatory discharge limit is a function of the duration of the release as given by the following equation:

$$TRC = [1074 \, (duration)^{-0.74}] \, \mu g/L,$$

where duration is the uninterrupted exposure period in minutes.

For intermittent flows, total duration of exposure in any consecutive 24-hour period should not exceed 2 hours. In addition, the maximum concentration of total residual chlorine should not exceed 100 $\mu g/L$ (0.1 mg/L) regardless of the exposure period. Allowable TRC concentrations for select duration of exposures are shown in Table 2.6.

Various other provincial Acts also facilitate implementation of WQS in BC. The Environmental Management Act confers broad duties, powers and functions upon the Environment Minister in matters relating to the management, protection, and enhancement of the environment including water resources. The Waste Management Act is designed to prohibit the introduction of waste in the environment in such a manner or quantity to cause pollution. The B.C. Water Act provides for the issuance of water licenses to divert and use or store water, or to undertake works in a stream or channel as provided in the license.

Manitoba

The Ministry of Environment in Manitoba uses Water Quality Objectives based on USEPA criteria to regulate discharges. This applies to discharges from drinking water sources also. However, the province does not currently have a general permit to regulate discharges from potable water sources. Also, the province has not specifically issued any individual permit for release of chlorinated waters. Backwash waters are generally sent to storm water retention tanks for chlorine dissipation or discharged to sanitary sewers. The agency is considering developing a general permit for chlorinated water discharges in the future.

Table 2.6

Approved and working criteria for water quality, British Columbia Ministry of Environment, Lands and Parks (1995)

Duration of exposure (minutes)	Allowable TRC in freshwater (mg/L)	Allowable CPO* in marine and estuarine waters (mg/L)
≤ 0.2	0.1	0.04
≤ 25	0.1	NA†
30	0.087	0.005
90	0.038	0.003
> 120 continuous criterion	0.002	0.003

Source: Greater Vancouver Regional District, 1997, *Construction Water Use Guidelines*

* Chlorine produced oxidant – Approved & Water Quality Criteria, Ministry of Environment, Land and Parks (1995).
† Not Available

New Brunswick

New Brunswick does not have a general permit for disposal of chlorinated water from potable water sources. Utilities must obtain individual permits to release chlorinated water from various activities. Utilities are recommended to discharge chlorinated waters in sanitary sewers or they are required to dechlorinate the water prior to discharge in storm sewers or streams.

Newfoundland

Newfoundland controls discharges of chlorinated water using the Environmental Control Water and Sewage Regulations under the Environment Act (O.C. 96-254). The chlorine discharge limit under this program is 1 mg/L. The discharge water pH must be between 5.5 and 9.0. Currently, no permit program is in place for potable water discharges.

Nova Scotia

The Ministry of Environment currently has a general permit for construction of new water mains. Under this permit all chlorine containing waters need to be completely dechlorinated (0.0 mg/L) prior to discharge into receiving streams. This permit program applies to all new and upgraded distribution systems, treatment plants and storage tanks. In addition, parties discharging chlorinated water must comply with Federal receiving water quality criteria for chlorine and other contaminants. Nova Scotia currently does not have a permit program for main flushing and existing main repair works. However, a new permit program is currently being drafted which is comprehensive and covers all chlorinated water discharges.

Ontario

The Ministry of Environment is responsible for maintaining water quality in the province. The Ministry of Environment implements water quality criteria in Ontario under the Ontario Water Resources Act. The Ministry considers all discharges to surface waters from potable water sources as wastewaters. A standard permit (not a general permit) is issued to larger utilities discharging backwash, main flushing and chlorinated waters from operations and maintenance activities. WQ Standards (0.002 mg/L) are to be maintained during all discharges. Individual permits for less frequent releases also are issued.

Prince Edward Island

Prince Edward Island (PEI) primarily uses groundwater for drinking water. The quality of groundwater in PEI is so high that no treatment, including chlorination, is currently in place. Hence, discharge of filter backwash is not an issue at PEI. When distribution mains are extended or repaired or newly laid, utilities get individual permits to discharge chlorinated water releases. No numerical limit for chlorine is specified, but all these waters must be dechlorinated prior to discharge into receiving streams. Utilities are required to ensure that no deleterious effect to aquatic species occurs due to release of such waters.

Saskatchewan

Saskatchewan Ministry of Environment has a receiving water quality objective for various contaminants. For chlorine, the concentration is 0.002 mg/L. Permissible total ammonia concentrations vary with pH and temperature. Total ammonia concentrations in the receiving water can be as high as 2.6 mg/L at pH 6.0 and 0 °C to as low as 0.06 mg/L at pH 9.0 and 30 °C. The warm water quality objective for dissolved oxygen varies from 5.0 to 6.0 mg/L and the cold-water quality criterion varies between 6.5 to 9.5 mg/L.

Saskatchewan has a general permit for the release of hydrotesting waters. Permissible chlorine concentrations are decided on a case-by-case basis, in accordance with WQS. Permit programs are not in place for other chlorinated water releases.

Quebec

The Quebec Ministry of Natural Resources decides allowable chlorine discharge level. A general permit for chlorinated water discharge is not currently available. The department assesses river stream conditions (e.g. type of fish species present, river flow, sensitivity, etc.) and issues permit on a case-by-case basis. In some cases of large rivers (such as St. Lawrence River), chlorine levels as high as found in drinking water may be permitted.

In Quebec, chlorination of wastewater effluents is not permitted. Wastewater discharges are often disinfected using UV light. Hence, no regulations exist for dechlorination wastewater releases.

Summary of Canadian Regulations

Table 2.7 summarizes the water quality criteria and permissible discharge chlorine concentrations in various provinces.

Table 2.7

Provincial regulatory information pertaining to effluent chlorine concentration

Province	Chlorine regulation
Alberta	Water Quality Criterion of 0.002 mg/L for the receiving streams.
British Columbia	Water Quality Criterion of 0.002 mg/L for the receiving streams. Concentration in the discharge water may vary from 0.002 to 0.1 mg/L (for continuous and intermittent flows).
Manitoba	Water Quality Criterion of 0.002 mg/L for the receiving streams.
New Brunswick	Water Quality Criterion of 0.002 mg/L for the receiving streams.
Newfoundland	Water Quality Criterion of 0.002 mg/L for the receiving streams. Chlorine concentration up to 1 mg/L in discharge waters.
Nova Scotia	Water Quality Criterion of 0.002 mg/L for the receiving streams.
Ontario	Water Quality Criterion of 0.002 mg/L for the receiving streams.
Prince Edward Island	Water Quality Criterion of 0.002 mg/L for the receiving streams.
Saskatchewan	Water Quality Criterion of 0.002 mg/L for the receiving streams.
Quebec	Water Quality Criterion of 0.002 mg/L for the receiving streams.

SUMMARY

A review of state and provincial regulations, and permit programs yields information on regional dechlorination practices. In general, discharge limits for most states and provinces are defined by receiving water quality criteria. In the U.S., the concentrations are 19 µg/L and 11 µg/L under acute and chronic toxicity criteria, respectively. In Canada, this concentration is 2 µg/L for receiving waters. State and provincial regulatory agencies require utilities to maintain these receiving water criteria during discharge of chlorinated waters from potable water sources.

However, some differences are observed in the approach to regulate potable water releases through NPDES permit programs. Among the western states in the U.S., California,

Oregon and Nevada have stringent regulatory discharge limits for chlorinated waters. Chlorine discharge limits in all water released to receiving streams must not exceed 0.1 mg/L (or a more stringent limit) in these states. Similarly, the states of Maryland and West Virginia in the eastern U.S. have a general permit for all chlorinated water releases. Chlorine discharge concentrations should be below 0.1 mg/L in these states also. Colorado, Connecticut, Tennessee, Kentucky, Wisconsin and Wyoming have more than one general permit to regulate various chlorinated water releases; however, these permits do not include all potable water discharges. Nebraska and Texas regulate hydrotesting waters through a general permit or an administrative rule. Utah has administrative guidelines for chlorinated water discharges. In many midwestern states like Iowa, Kansas, Illinois and Michigan and some southeastern states like Arkansas, Georgia and Louisiana, no general or individual permit program is in place for potable water releases. However, utilities in these states are required to meet the water quality criteria of receiving streams while discharging potable waters.

In Canada, British Columbia and Ontario require all water releases to contain less than 0.002 mg/L of chlorine. This concentration of chlorine is difficult to measure, even in the laboratory, using sophisticated equipment. GVRD recommends the use of pocket colorimeters for field measurement of residual chlorine. This technique is sufficiently sensitive to measure residual chlorine concentrations of as low as 0.1 mg/L. The generic emergency response plan developed by GVRD indicates that residual chlorine concentrations as low as 0.03 mg/L can be detected, but not quantified, by a slight change in color, using pocket colorimeters. In addition, GVRD recommends that utilities look for and record any observations of dead or stressed fish, impact to insects and other invertebrates, or other evidence of environmental damage to ensure compliance with residual chlorine limits.

The provinces of Nova Scotia and Manitoba are currently revising their permit program. The revised permit is expected to have stringent discharge concentrations for chlorine in potable water releases. Alberta does not have a general permit for potable water releases, but requires all utilities to meet the discharge limit of 0.002 mg/L through water treatment plant operating permits. Saskatchewan has a general permit for hydrotesting waters and decides chlorine discharge limits on a case-by-case basis. Newfoundland has a chlorine discharge limit of 1 mg/L.

Most states and provinces follow the USEPA regulatory criterion for ammonia. Dissolved oxygen concentration requirements for most states and provinces vary from 5 to 7 mg/L. The pH levels of the receiving water must be between 6.0 and 9.0 in most of the states and provinces.

CHAPTER 3

TYPES OF CHLORINATED WATER RELEASES

INTRODUCTION

Chlorinated water is released into the environment from a variety of distribution and water treatment plant related activities. Chlorine is present in these waters as a result of disinfection of system components (water mains, storage facilities, treatment plant, etc.) or disinfection of the potable water. The concentration of chlorine present and the amount of water released depend on the type of the water release.

The American Water Works Association (AWWA) has provided standards for the disinfection of water mains, storage facilities and treatment plants (AWWA Standards C651-92, C652-92 and C653-97, respectively). These standards provide a basis for estimating the concentration of chlorine in waters released by these types of activities. The Safe Drinking Water Act (SDWA) and relevant state drinking water regulations set minimum residual chlorine concentrations in potable waters. Chlorine concentrations in discharges from water distribution systems will be reflective of the practices of the utilities in meeting drinking water regulations. A summary of disinfection requirements prescribed by AWWA is presented in Table 3.1.

This report contains information obtained from the participating utilities and PAC member utilities regarding activities causing the release of chlorinated water. Information related to volume, rate and chlorine concentration is also included. These utilities differ in their use of disinfection agents (eight of the utilities use free chlorine and five use combined chlorine) and source waters (seven utilities primarily use surface water, one uses only groundwater and the rest use a combination of ground and surface waters). In addition, the population served by these utilities is also significantly different (110,000 to 1,800,000). Some utilities wholesale most of their waters and the rest of the utilities retail most of their waters. All these differences have an impact on the water management practices and hence in the quantity and character of the chlorinated water released.

Table 3.1

Summary of AWWA disinfection requirements

Component/activity	Suggested disinfection method	Chlorine-concentration (mg/L)
New main - wet trench water	Hypochlorite granules	25
New main - disinfection due to accidental flooding	Holding chlorinated water for 24 hours	25
New main disinfection	Tablets	25
	Continuous-feed (holding for 24 hours)	≥ 10
	Slug method (3 to 4 hour exposure)	≥ 50
Repair of existing mains	Slug method (3 to 4 hour exposure)	300
Water storage facilities	Holding chlorinated water for 24 hours	≥ 10
	Spraying surfaces with 200 mg/L Chlorine and purge with potable water after 30 minutes	≥ 10
	Two stage chlorine addition	50 (Residual Cl_2 - 2 mg/L)
Water treatment plant	Holding chlorinated water for 24 hours	25
	Chlorination during filter to waste	25

Reference: Adapted from American Water Works Association (AWWA) Standard for Disinfecting Water Mains (AWWA C651-92), AWWA Standard for Disinfection of Water-Storage Facilities (AWWA C652-92) and AWWA Standard for Disinfection of Water Treatment Plants (AWWA C653-97).

Table 3.2 provides a collective summary of the types of chlorinated water releases from various operation/maintenance practices that will be addressed by this manual. Generally, the types of releases can be categorized into planned, unplanned and emergency releases.

PLANNED RELEASES

Planned releases of chlorinated water result from operation and maintenance activities such as disinfection of mains, testing of hydrants/water mains, and routine flushing of distribution systems for maintenance. The volume, duration and chlorine concentrations vary with the type of activity. For example, following repair works, water mains are disinfected with highly chlorinated waters (200-300 mg/L), whereas extended disinfection (24 hour) of storage tanks use a lower chlorine concentration (10 mg/L). Waters released from flushing activities generally contain less than 4 mg/L. In addition, the released water characteristics vary with the policies and practices of the utility. For example, for routine water main flushing East Bay Municipal Utility District (EBMUD) uses low-flow (300 to 500 gpm), short duration (10 to 30 minutes) flushing, whereas Cincinnati Water Works (CWW) uses a much higher flow rate (2000 gpm) for 20 minutes (short duration) for such activities.

Although planned releases sometimes may contain high concentrations of chlorine, their discharges are easier to control and hence, easier to dechlorinate.

UNPLANNED RELEASES

Unplanned releases occur from activities such as main breaks, leaks, overflows and emergency flushing activities. Unplanned releases in most cases typically have a lower chlorine concentration (reflective of chlorine concentrations in the distribution system) than concentrations in certain planned releases. However, they are harder to neutralize due to limitations in response time, staff availability and the difficulty in containing these waters. In addition, during unplanned release events such as water main breaks, higher priority is given to protecting public health, shutting off or reducing flow and restoring service to customers. The duration of unplanned releases varies with the type of activity. A collective summary of the type, volume, duration, chlorine concentrations of different planned and unplanned chlorine releases is presented in Table 3.2.

EMERGENCY RELEASES

Activities such as water main flushing in response to water quality complaints from the public, and releases during fire fighting are examples of emergency releases of chlorinated waters. The amount of water released and the chlorine concentrations vary significantly with the type of release. Emergency releases of chlorinated waters are the most difficult types to dechlorinate, because of the extremely low predictability of these events and limitations in the response time.

Table 3.2

Types of potable water discharges containing chlorinated water

Discharge activity	Typical flow rate range (gpm)	Typical discharge duration (minutes)	Typical max Cl_2 residual (mg/L)	Type of release
Main dewatering				
- for new construction (by utility)	200-500	120-720	2.5	planned
- for new construction (contractor)	"	"	2	planned
- for maintenance	"	"	0.8-2	both
- large pipe inspection	2000	1-2 days	0.8-2	planned
Pumping plant and reservoir O&M activities				
- new construction (disinfection)	200-400	varies	10-50	planned
- maintenance or construction related (in-house)	5-200	2-4 min	0.8-2	both
- drain valve testing	5-100	60-120	2	planned
- reservoir overflow	varies	< 120	2	unplanned

(continued)

Table 3.2 (Continued)

Discharge activity	Typical flow rate range (gpm)	Typical discharge duration (minutes)	Typical max Cl_2 residual (mg/L)	Type of release
- underground emergency scenario	varies	varies	2	unplanned
- dead-end pumping to relieve excess	200-1000	varies	2	both
- reservoir rehab pipe flushing	varies	varies	2	planned
- tank/reservoir draining for maintenance	varies	varies	0.8-2	planned
- tank freshening	5-30 GPM		0.8-2	planned
Main flushing and service line flushing				
- in response to taste & odor complaints	100-1000	10-60	0.8-2	emergency
- in response to total coliform rule exceedance	300-1000	10-60	0.8-2	emergency
- following disinfection (by utility)	300-2000	10-120	0.8-200	planned
- following disinfection (by contractor)	300	10-30	50-200	planned
- preventative (to avoid water quality concerns)	300-1000	10-60	0.8-2	planned
- emergency flushing (for public health concerns under emergency circumstances)	varies	varies	0.8-2	emergency
- new cement-lined pipes	30-60	1-2 days	0.8-2	planned
- temporary by-pass lines	20	24	0.8-2	planned
Standpipe cleaning	500-2000	1-2 days	0.8-2	planned
Field testing of water meters	50-1000	30-60	0.8-2.5	planned
Hydrant testing	700-1600	≤ 5	0.8-2	planned
Fire fighting	Varies	Varies	0-2	emergency

(continued)

Table 3.2 (Continued)

Discharge activity	Typical flow rate range (gpm)	Typical discharge duration (minutes)	Typical max Cl_2 residual (mg/L)	Type of release
Planned distribution system maintenance (trench dewatering)	5-200	10-60	0.8-2	planned
Unauthorized hydrant opening	500-1000	60-480	0.8-2	unplanned
Water main breaks (trench dewatering)	5-200	30-180	0.8-2	unplanned
Aqueduct dewatering				
- by utility	250-50,000	1-2 days	0.8-2	planned
- by contractor	250-500	1-2 days	0.8-2	planned
- aqueduct release due to high pressure	250-50,000	1-2 days	0.8-2	unplanned
Leakage				
- from reservoir altitude valves	0.5-1	varies	0.8-2	unplanned
- from reservoir underdrains	1-50	continuous	0.8-2	unplanned
- from treatment plant basins	1-100	continuous	0-2	unplanned
- from temporary bypass lines	1-50	30-90 minutes	0.8-2	unplanned
Treatment plant operational release				
- Filter to waste	2-5 gpm/ft^2	20 minutes	0.5	planned
- Filter backwash	15-20 gpm/ft^2	15 minutes	0.5	planned
- Plant overflow	varies	< 120	2-5	unplanned
- Sludge/water from sedimentation basins	500-2000	varies	0.5-1.0	planned
Treatment plant new construction/modification				
- Plant disinfection	varies	varies	10-50	planned
- Draining of plants and clear wells	varies	varies	0.8-5	planned
- On site water main breaks	5-200	30-180	0.8-5	unplanned

SUMMARY

Table 3.3 lists chlorinated water releases originating from various operation and maintenance activities. In general, the approach to disposal of chlorinated water depends on the amount of chlorine present, volume and flow rate of water to be neutralized and response time available for planning. Hence, these releases are grouped into the following categories:

Table 3.3

Categories of chlorinated water releases

Type of release	Chlorine concentration	Volume or flow rate
Planned	Low chlorine	High flow rate
Planned	Low chlorine	Moderate flow rates
Planned	Low chlorine	Low flow rates
Planned	High chlorine	Moderate flow rates
Unplanned	Low chlorine	High flow rates
Unplanned	Low chlorine	Moderate flow rates
Unplanned	Low chlorine	Low flow rates

In developing these categories, low chlorine releases are defined as those containing less than 4 mg/L chlorine. Flows greater than 500 gpm and lasting for more than a day are defined as high flows, and flows smaller than 50 gpm and lasting for less than 2 hours are classified as low flow releases. Remaining flows are moderate flow releases. Such categorization of releases may help in development of guidance and BMPs for dechlorination, once sufficient field dechlorination data is available. The releases under each category are summarized below.

1) Low chlorine, high flow release
   a. Main dewatering (> 24" diameter)
      i. Large pipe inspection
   b. Pumping plant and reservoir maintenance
      i. Tank/Reservoir draining for maintenance
   c. Standpipe cleaning

- d. Aqueduct dewatering
- e. Aqueduct release due to high pressure
- f. Treatment plant new construction modifications
    - i. Draining of plants and clearwells

2) Low chlorine, moderate flow releases
- a. Main dewatering (< 24" diameter)
    - i. For new construction
    - ii. For maintenance
- b. Pumping plant and reservoir operation and maintenance
    - i. New construction (disinfection)
    - ii. Drain valve testing
    - iii. Dead end pumping to relieve excess pressure
    - iv. Reservoir rehabilitation pipe flushing
- c. Main flushing
    - i. Pigging and swabbing
    - ii. In response to taste and odor concerns
    - iii. In response to Coliform Rule
    - iv. Preventative (to avoid water quality concerns)
    - v. Emergency flushing
    - vi. New cement lined pipes flushing
    - vii. Temporary by-pass line flushing
- d. Hydrant testing
- e. Planned distribution system maintenance (trench dewatering)
- f. Treatment plant operational releases
    - i. Filter to waste
    - ii. Filter backwash
    - iii. Sludge/water from sedimentation basins

3) Low chlorine, low flow releases
- a. Pumping plant and reservoir operation and maintenance activities
    - i. Maintenance or construction related
    - ii. Tank freshening

4) High chlorine, moderate flow releases
   a. Pumping plant and reservoir operation and maintenance activities
      i. New construction disinfection
   b. Main flushing
      i. Following disinfection
   c. Treatment plant new construction/modification
      i. Plant disinfection
5) Unplanned, low chlorine, moderate flow releases
   a. Pumping plant, reservoir operation and maintenance activities
      i. Reservoir overflow
   b. Water main breaks (smaller branches)
   c. Unauthorized hydrant opening
6) Unplanned, low chlorine, high flow releases
   a. Pumping plant, reservoir activities
      i. Underground emergency scenarios
   b. Water main breaks
7) Unplanned low chlorine, low flow releases
   a. Leakages
      i. From reservoir altitude valves
      ii. From reservoir underdrains
      iii. From treatment plant basins
      iv. From temporary by-pass lines

CHAPTER 4

CURRENT DISPOSAL PRACTICES

INTRODUCTION

Chlorine is a relatively unstable, moderately reactive element. In the environment, chlorine is neutralized upon reaction with air, sunlight and other contacting surfaces. Furthermore, chlorine readily reacts with organic and inorganic impurities in soil, paved surfaces, water and wastewater. For example, chlorine reacts with hydrogen sulfide to form hydrochloric acid, and is rapidly neutralized by reducing agents such as Fe^{2+}, Mn^{2+} and NO_2^- in water and wastewater. Chlorine also reacts with organic compounds having unsaturated linkages, such as humic substances, to form a variety of organic-chlorine compounds. Hence, many utilities currently dispose chlorinated water by retention in holding tanks and infiltration ponds, discharge into sanitary sewers, and release to soil surfaces.

Chlorine can also be chemically neutralized by reaction with various dechlorination agents such as sulfur dioxide, sodium/calcium thiosulfate, sodium sulfite, sodium bisulfite, and sodium metabisulfite. Chlorine can also be removed by activated carbon.

The persistence of chlorine in the environment, however, depends on the form of chlorine present. Utilities often use free or combined chlorine (chloramine) for disinfection of potable water. Free chlorine is applied as chlorine gas or sodium/calcium hypochlorite. In combined chlorine systems, in addition to chlorine, ammonia is added to the water at a chlorine-to-ammonia ratio of 3:1 to 6:1 (White 1999).

Free chlorine is a stronger disinfectant and reacts more rapidly in the environment than combined chlorine. Chloramines are very persistent and their reactions with air, sunlight, organic and inorganic compounds are much slower.

In this section, the effectiveness of various passive non-chemical methods for disposal of chlorinated water is discussed. Furthermore, dechlorination properties of chemical agents currently used, related water quality impacts, health and safety concerns, dose calculations and costs are also presented. Treatment of both free and combined chlorine using these techniques is discussed.

This section also summarizes various flow control measures for the release of chlorinated waters. Techniques currently used to introduce dechlorination agents into the discharge are also presented. Dechlorination techniques currently practiced by participating utilities, PAC member utilities, selected other utilities and industries are also presented.

NON-CHEMICAL METHODS FOR CHLORINE DISSIPATION

Retention in Holding Tanks

The chlorine concentration in stored water gradually decreases with time due to aeration, reaction with sunlight and reaction with surfaces of holding tanks. Several utilities in the United States and Canada store filter backwash water and main disinfection water in holding tanks to allow for residual chlorine decay prior to discharge. Since chemical neutralization of chlorine from super-chlorinated water requires a large amount of the dechlorination agent, some utilities reduce chlorine concentrations in these waters by storage for an extended period in holding tanks, prior to adding dechlorinating chemicals.

The advantage of dissipating chlorine passively is that such a process does not involve chemical addition. Hence utilities do not have to be concerned with the effects of neutralizing chemicals in the receiving streams. Also avoided are cost, health and safety concerns related to storage, transportation and handling of these chemicals.

However, this method of dechlorination has several limitations. First, chlorine decay through natural reactions is extremely slow. Several studies have been performed to evaluate natural chlorine decay rate in bulk solutions and distribution systems. Free chlorine decay under these conditions is reported to be a first order reaction with a decay coefficient varying from 0.85 to 0.1 day^{-1} (Vasconcelos et al. 1997). These values indicate that decay of free chlorine at concentrations typically found in distribution systems (0.5 to 2 mg/L) will take several hours to a few days to meet regulatory discharge limits (0.1 to 0.002 mg/L). Second, activities such as reservoir cleaning and large main dewatering produce a large volume of chlorinated water, requiring very large tanks for storage. Also, it may be difficult and expensive to transport holding tanks to various dechlorination sites in the distribution system. Finally, transport of holding tanks may be difficult and expensive, while responding to unplanned and emergency release of chlorinated waters.

Combined chlorine is more stable than free chlorine in the environment. Decay rates are reported to be three to fourfold slower than that of free chlorine. Hence, decay of combined chlorine in holding tanks will require much longer retention times than those of free chlorine.

Land Application of Chlorinated Water

Organic and inorganic impurities in soil and pavements exert a significant amount of chlorine demand and rapidly neutralize chlorine in waters. Hence, spraying chlorinated waters onto soils or pavements can be a very effective method for disposing of chlorine-containing waters. This method also does not introduce dechlorinating chemicals into the water. The presence of impurities in the soil typically increases decay rates compared to those observed in holding tanks.

However, potential drainage of waters applied on land, particularly from recently cleaned roads and pavements, into storm drains and receiving waters is a matter of concern while using this method. Also, land application of large volumes of water may lead to soil erosion. Since chloramines are more stable than free chlorine, a longer travel/detention time is required for the dissipation of this disinfectant.

EBMUD conducted preliminary empirical studies in early 1998 to evaluate the decay of combined chlorine when discharged onto different surfaces. Chloraminated waters containing 1 - 2 mg/L of total residual chlorine were released at 300 - 500 gpm, as sheet flow, onto dirty gravel or pavement surfaces on a sunny day. At the highest flow rate, the discharged traveled approximately half a mile (2,414 feet) in 28 minutes. The tests did not conclusively determine the distance from the point of release where the chlorine residual is reduced to below detection limits. However, the studies indicated that water had to travel at least half a mile prior to decay below regulatory discharge concentrations. A study was conducted at Tacoma Waters facility to evaluate decay of free chlorine on paved surfaces. Water containing 1.2 mg/L free chlorine was released from a hydrant at 300 gpm, as sheet flow on a semi-paved surface. The flow traveled

500 feet in 4 minutes and 10 seconds and only about 0.2 mg/L of free chlorine was dissipated due to chlorine demand of the surface. Hence, this technique appears to be more effective for discharging small amounts of water, in locations far from storm drainage and receiving streams.

Discharge of Chlorinated Water for Groundwater Recharge

The Metropolitan Water District of Southern California (MWD) sometimes discharges chlorinated water to dry streambeds or to land for groundwater recharge. Hay bales may have to be keyed into the soil and staked to avoid water running under the bales. This is an acceptable dechlorination practice if the water percolates before it reaches another body of water. Currently, no standard practices have been developed for this activity. However, prior to any discharge, MWD always surveys the area where the discharge will go and estimates how far it will travel based upon the quantity and discharge rate. MWD also coordinates these efforts with the local flood control entity, where appropriate. During disposal of chlorinated water for groundwater recharge, MWD always has dechlorination equipment available for use if necessary.

Discharging Through Hay Bales and Other Natural Obstructions

Backwash and planned water releases from the distribution system may be allowed to flow through hay bales or other obstructions to dissipate chlorine prior to discharging into storm sewers and receiving waters. While the chlorine demand exerted by these obstructions can be reasonably high, it may be difficult to achieve regulatory discharge limits in some cases. Also, elaborate arrangements required to construct such barriers, practical difficulties in construction of such barriers at various field discharge points, and potential soil erosion while discharging waters are some of the concerns in using this technique.

Discharge to Storm Sewers

Discharging chlorinated water into storm sewers may be an effective way to dissipate chlorine from some potable water releases. However, there are several limitations and risks in releasing chlorinated waters into storm drains. Although storm waters may contain some organic and inorganic impurities, their concentrations may not be sufficient to completely dechlorinate the water released. Lower chlorine demand in storm waters may require a longer travel time for chlorine neutralization. Storm waters are usually discharged directly into receiving streams or waters leading to aquatic species bearing streams. In some instances, therefore, chlorinated waters released into storm sewers may not undergo sufficient dechlorination before being discharged to nearby receiving streams.

Discharge of Chlorinated Waters in Sanitary Sewers

The release of chlorinated water into sanitary sewers is a very safe and effective means of disposing chlorinated waters in most cases. Most of the water utilities prefer this method as their first option for releasing chlorinated potable waters. Many utilities prefer to discharge superchlorinated water into sanitary sewers. However, this requires close coordination with sanitation district officials to minimize any adverse impact to the sewage treatment plant operations. A very high demand exerted by sulfide and other inorganic and organic pollutants in sewage rapidly neutralizes chlorine. In addition, since the water is not directly released to receiving streams, utilities do not have to be concerned with meeting receiving WQS and discharge limits. Also, this non-chemical method of neutralization eliminates potential water quality concerns caused by dechlorination agents.

The availability of a sanitary sewer near the point of chlorinated water release, and the capacities of the sanitary sewer and the wastewater treatment plant to handle the additional load are the primary limitations in this method. Potential upset of treatment plant operations due to chlorinated water release must also be evaluated. Permission and coordination from the responsible sanitation authority must be obtained prior to releasing chlorinated water into sanitary sewers to ensure safety of workers who may be working in the sewer lines. Utilities may be required to pay a discharge fee to the sanitation district for releasing chlorinated water into sanitary sewer. Finally, caution must be exercised in avoiding potential cross-connection during discharge to sanitary sewers. A backflow prevention device or an air gap method must be used to prevent cross-connection problems.

Dechlorination Using Activated Carbon

Activated carbon has been widely used in water and wastewater treatment facilities and industries for dechlorination (White 1999). Studies indicate that activated carbon can remove free as well as combined chlorine from water. Granular activated carbon (GAC) is often used for dechlorination activities. In water treatment plants, carbon filters effectively remove dissolved organic matter in addition to removing chlorine.

Free chlorine is removed in carbon filters by the following reactions:

$$C^* + HOCl \rightarrow CO^* + HCl$$
Active carbon — Hypochlorous acid — Surface oxide on carbon — Hydrochloric acid

If significant amount of chlorine reacts with carbon, some of the oxygen attached with carbon may emit as CO or CO_2 gas (White 1999).

$$C + 2Cl_2 + 2H_2O \rightarrow 4HCl + CO_2$$
Carbon — Chlorine — Hydrochloric acid — Carbon dioxide

A fraction of carbon is permanently destroyed during this reaction. According to stoichiometry, one part of chlorine can destroy 0.00845 parts of carbon by this reaction.

Carbon undergoes the following reactions with mono and dichloramines:

$$2NH_2Cl + H_2O + C^* \rightarrow NH_3 + HCl + CO^*$$
Monochloramine — Carbon — Ammonia — Hydrochloric acid — Surface oxide on carbon

$$2NHCl_2 + H_2O + C^* \rightarrow N_2 + 4HCl + CO^*$$
Dichloramine — Carbon — Nitrogen — Hydrochloric acid — Surface oxide on carbon

Unlike free chlorine reactions, chloramines reactions do not release CO_2 gas. Hence, the carbon is not destroyed during dechlorination of chloramines. However, chloramines reactions are much slower than chlorine reactions with activated carbon.

Although carbon can effectively remove chlorine from potable water, several studies indicated that it is more expensive than other dechlorination methods (White 1999; Metcalf and Eddy 1981). In addition, application of carbon for dechlorination activities is not studied extensively.

DECHLORINATION USING CHEMICALS

Whenever it is not possible to dispose of chlorinated waters safely by non-chemical methods, chlorine must be neutralized using dechlorination chemicals. Several solid, liquid and gaseous dechlorination chemicals are commercially available and are widely used by water and wastewater utilities. This section describes the reactions of various chemicals with free and combined chlorine, related water quality and health and safety issues, ease of use, cost and other issues related to the application of these chemicals.

Sulfur Dioxide (SO_2)

Sulfur dioxide is a colorless gas with a suffocating pungent odor. It is widely used in water and wastewater treatment plants for dechlorinating backwash water and wastewater disinfected with chlorine. Sulfite ion is the active agent when sulfur dioxide is dissolved in water. Sulfite reacts instantaneously with free and combined chlorine according to the following stoichiometry (Snoeyink and Suidan 1975):

$$SO_2 + H_2O + HOCl \rightarrow SO_4^{-2} + Cl^- + 3H^+$$
Sulfur dioxide / Hypochlorous acid / Sulfate

$$SO_2 + NH_2Cl + H_2O \rightarrow SO_4^{-2} + Cl^- + NH_4^+ + 2H^+$$
Sulfur dioxide / Monochloramine / Sulfate ammonia

As shown, the reaction of sulfur dioxide with free chlorine produces hydrochloric acid and reaction with combined chlorine produces ammonium chloride.

Effect on Water Quality

Dechlorination using sulfur dioxide produces a small amount of acid. Approximately 2.8 mg of alkalinity as $CaCO_3$ is consumed per milligram of chlorine reduced. SO_2 is also an oxygen scavenger. It can deplete dissolved oxygen in the discharge water and in the receiving stream.

Dosage Calculations

The mass ratio of SO_2 to available chlorine is 0.9:1. In the field, nearly 1.1 parts of SO_2 are required to neutralize 1 part of chlorine (WPCF 1986).

Health and Safety Issues

Sulfur dioxide is a toxic chemical subject to reporting requirements of the Superfund Amendments and Reauthorization Act (SARA), Title III, Section 313. It has a National Fire Protection Association (NFPA) rating of 2, 0 and 0 for health, fire and reactivity, respectively. (Hazard rating is from 0 to 4, with 0 indicating no hazard and 4 indicating extremely hazardous). It is an extremely irritating gas. It may cause various degrees of irritation to mucous membranes of the eyes, nose, throat and lungs. Contact with sulfur dioxide liquid may produce freezing of the skin because the liquid absorbs the heat of vaporization from the skin. Concentrations above 500 mg/L can cause acute irritation to the upper respiratory system and cause a sense of suffocation. The Threshold Limit Value, Time Weighted Average (TLV:TWA) is 2 ppm and the Short Term Exposure Limit (STEL) value of SO_2 is 5 ppm. Great caution is required in transporting cylinders of SO_2 gas.

Stability and Reactivity

SO_2 is a stable compound. It reacts with water to produce sulfurous acid.

Other Dechlorination Issues

SO_2 is a hazardous gas. It must be stored, transported and handled with care. It is a toxic chemical subject to reporting requirements of SARA, Title III, Section 313. Care must be taken to design storage and handling facilities to avoid accidental exposure to gas release. While it is suitable for use in facilities such as treatment plants and pumping stations, it is not best suited for field applications. Hence, SO_2 will not be considered for dechlorination of various potable water releases in this report.

Sodium Thiosulfate ($Na_2S_2O_3$)

Sodium thiosulfate is a colorless, transparent monoclinic crystal widely used by utilities for dechlorination. It undergoes various reactions with free and combined chlorine, depending on solution pH (Snoeyink and Suidan 1975; General Chemical 1988). Reaction with chlorine yields the following:

$$Na_2S_2O_3 + 4HOCl + H_2O \rightarrow 2NaHSO_4 + 4HCl$$
Sodium thiosulfate, Hypochlorous acid, Sodium bisulfate, Hydrochloric acid

$$Na_2S_2O_3 + HOCl \rightarrow Na_2SO_4 + S + HCl$$
Sodium thiosulfate, Hypochlorous acid, Sodium sulfate, Hydrochloric acid

$$2Na_2SO_3 + HOCl \rightarrow Na_2S_4O_6 + NaCl + NaOH$$
Sodium thiosulfate, Hypochlorous acid, Sodium tetrathionate, Sodium chloride, Sodium hydroxide

Sodium thiosulfate undergoes the following reactions with monochloramines:

$$4 NH_2Cl + Na_2S_2O_3 + 5 H_2O \rightarrow 2 NaHSO_4 + 4 NH_3 + 4 HCl$$
Monochloramine, Sodium thiosulfate, Sodium bisulfate, Ammonia, Hydrochloric acid

$$NH_2Cl + Na_2S_2O_3 + H_2O \rightarrow Na_2SO_4 + S + NH_3 + HCl$$
Monochloramine, Sodium thiosulfate, Sodium Sulfate, Ammonia, Hydrochloric acid

$$NH_2Cl + 2 Na_2S_2O_3 + 2 H_2O \rightarrow Na_2S_4O_6 + 2 NaOH + NH_3 + HCl$$
Monochloramine, Sodium thiosulfate, sodium tetrathionate, Sodium hydroxide, ammonia, Hydrochloric acid

Sodium thiosulfate undergoes the following reactions with dichloramines

$$2NHCl_2 + Na_2S_2O_3 + 5 H_2O \rightarrow 2 NaHSO_4 + 2 NH_3 + 4 HCl$$
Dichloramine, Sodium thiosulfate, Sodium bisulfate, Ammonia, Hydrochloric acid

$$NHCl_2 + 2 Na_2S_2O_3 + 2 H_2O \rightarrow 2 Na_2SO_4 + 2 S + NH_3 + 2 HCl$$
Dichloramine, Sodium thiosulfate, Sodium Sulfate, Ammonia, Hydrochloric acid

$$NHCl_2 + 4 Na_2S_2O_3 + 4 H_2O \rightarrow 2 Na_2S_4O_6 + 4 NaOH + NH_3 + 2 HCl$$
Dichloramine, Sodium thiosulfate, Sodium tetrathionate, Sodium hydroxide, ammonia, Hydrochloric acid

Dechlorination of chloramines may produce NH_3 or NH_4^+, depending on pH. The reaction produces sodium bisulfate, sodium sulfate and/or HCl. In addition, chloramine reactions produce trace concentrations of ammonium chloride.

Effect on Water Quality

The pH of sodium thiosulfate solution is near neutral. Although it produces HCl during dechlorination, studies performed by EBMUD in 1998 indicated that the reaction does not alter solution pH appreciably, under the conditions studied. In these studies, flow from a hydrant was released at a rate of 100 gpm and a bag containing 1 lb of sodium thiosulfate was placed in the path of the released water. The storm water inlet was located approximately 160 feet away from the point of release. Results indicated that sodium thiosulfate reduced chlorine concentrations from approximately 1 mg/L to less than 0.1 mg/L in the released water. Although the pH of the released water momentarily decreased from 8.9 to 8.6, the pH subsequently increased to 8.9 during these studies. (Note: The storm water inlet discharged to the District's wastewater treatment plant, which eliminated any possibility of discharging chlorinated water to surface water during these studies).

Thiosulfate is an oxygen scavenger and reducing agent. However, it scavenges less oxygen than sodium sulfite, bisulfite or metabisulfite. In the above study using thiosulfate, the dissolved oxygen concentrations initially decreased from 9.5 mg/L to 8.6 mg/L. However, the concentrations increased to 9.5 mg/L after one minute of contact with the chemical.

Dosage Calculations

On a weight-to-weight basis, to neutralize one part of chlorine, the three reactions (yielding sodium bisulfate, sodium sulfate and sodium tetrathionate, respectively) require 0.556, 2.225 and 4.451 parts of sodium thiosulfate, respectively, (Snoeyink and Suidan 1975; General Chemical 1988). The amount of thiosulfate required is independent on the form of chlorine (free or combined chlorine) present. However, the amount of thiosulfate required for dechlorination may vary with solution pH (General Chemical 1988). Approximately 2.23 parts of thiosulfate are required to neutralize one part of chlorine at pH 6.5 and nearly 1.6 parts of sodium thiosulfate is sufficient to neutralize one part of chlorine at pH 9.0.

Health and Safety Issues

Sodium thiosulfate is a skin, eye, nose and throat irritant. It is moderately toxic by an intravenous route. It has a NFPA Rating of 1, 0, 0 for health, fire, and reactivity, respectively. (Hazard rating is from 0 to 4, with 0 indicating no hazard and 4 indicating extreme hazard). A USEPA (1988) toxicity study indicated that sodium thiosulfate is not very toxic to aquatic species. For ceriodaphnia, the 24 and 48 hour LC50 values are 2.5 and 0.85 g/L respectively. For daphnia, these values are 2.2 and 1.3 g/L, respectively. For fathead minnows, the LC50 values for 24, 48, 72 and 96 hours are 8.4, 8.4, 7.9 and 7.3 g/L, respectively.

Stability and Reactivity

No published data are available on the stability of sodium thiosulfate. However, utilities using thiosulfate for dechlorination have reported that the strength of thiosulfate solutions does not decrease appreciably after 2 or 3 days of storage.

Sodium thiosulfate reacts with acid to produce SO_2 and H_2S. It reacts violently with $NaNO_2$. When heated to decomposition, it emits toxic fumes of sulfur dioxide and Na_2O. It reacts rapidly with iron and is readily hydrolized by water (i.e., it is hygroscopic).

Costs

Costs typically range from $0.70 to $ 1.50 per pound (technical grade) for thiosulfate delivered in 50 to 100 pound containers. Actual costs will depend upon quantities ordered and delivery locations.

Other Issues

Sodium thiosulfate is not currently available in tablet form, which is easier to transport, store and use. However, many utilities prefer to use a sodium thiosulfate solution because it is a weaker oxygen scavenger than other dechlorinating agents such as sodium sulfite, bisulfite and metabisulfite. In addition, the solution allows for better dose control during dechlorination.

A concern over using sodium thiosulfate is that this chemical reacts slowly with chlorine and requires more time for dechlorination than sulfur dioxide and other dechlorination chemicals (White 1999). In addition, over-dechlorination with sodium thiosulfate may encourage thiobacillus and some other bacterial growth in receiving streams, particularly during low flow conditions. A drop in pH, caused by the production of H_2SO_4 by microorganisms, has been reported under such conditions.

Sodium Sulfite (Na_2SO_3)

Sodium sulfite is yet another dechlorinating agent widely used by utilities. It is generally available in powder/crystalline form. In addition, some companies (Exceltec and Norweco) produce sodium sulfite tablets. Sodium sulfite undergoes the following reaction with free chlorine (General Chemical 1988):

$$Na_2SO_3 + HOCl \rightarrow Na_2SO_4 + HCl$$
Sodium sulfite — Hypochlorous acid — Sodium sulfate — Hydrochloric Acid

Reactions with mono/dichloramines are as below:

$$Na_2SO_3 + NH_2Cl + H_2O \rightarrow Na_2SO_4 + NH_3 + HCl$$
Sodium sulfite — Monochloramine — Sodium sulfate — Ammonia — Hydrochloric acid

$$2\,Na_2SO_3 + NHCl_2 + 2\,H_2O \rightarrow 2\,Na_2SO_4 + NH_3 + 2\,HCl$$
Sodium sulfite — Dichloramine — Sodium sulfate — Ammonia — Hydrochloric acid

Sodium sulfite produces sodium sulfate and hydrochloric acid with free and combined chlorine. Ammonia may be present as NH_3 or NH_4^+, depending on pH. In addition, reaction with chloramines produces ammonium chloride.

Effect on Water Quality

Sodium sulfite solutions are slightly alkaline. Fifty grams of sodium sulfite (anhydrous) in one liter of distilled water produces a solution with a pH of 8.5-10.5 (Mallinckrodt Baker 1999). Although it produces HCl, field studies (Section 6) indicated that the reaction doesn't appreciably decrease solution pH during dechlorination.

Sodium sulfite is a reducing agent and is reported to scavenge more oxygen than sodium thiosulfate. About eight parts of sodium sulfite are required to neutralize one part of oxygen. However, field studies performed by EBMUD as well as the Exceltec Company indicated that sodium sulfite tablets removed less than 10% of dissolved oxygen from waters containing approximately 1 mg/L of residual chlorine.

Dosage Calculations

On a weight-to-weight basis, approximately 1.775 parts of sodium sulfite are required to remove one part of chlorine (Bean 1995).

Health and Safety Issues

Sodium sulfite may affect the brain, respiratory system, and skin. It is an eye, skin, mucous membrane and respiratory tract irritant. It has a hazard rating (NFPA) of 2,0,0 for health, fire and reactivity.

Stability and Reactivity

Material Safety Data Sheets (MSDSs) indicate crystalline sodium sulfite to be stable. However, sodium sulfite solution will decompose upon reaction with air to form SO_2 gas. The shelf life of D-Chlor sodium sulfite tablets is reported to be about one year for unopened pails, if stored properly. If opened, the shelf life for the tablets is about two months.

Sodium sulfite reacts strongly with acids to produce SO_2. It is a strong reducing agent and reacts with oxidants and decomposes on heating to produce SO_2.

Cost

Estimated between $0.50 and $0.90 per pound (technical grade), if delivered in 100-pound containers. A 45 lb. pail of sodium sulfite tablets is sold for $124.00. Actual costs will depend on quantities ordered and delivery locations.

Other Issues

The major advantage of using sodium sulfite is that, currently, it is the only dechlorination chemical commercially available in tablet form. Many utilities find dechlorination tablets easier to store, handle and apply as compared to dechlorination solutions or powders. Although sodium sulfite is a reducing agent and can scavenge oxygen, field studies performed by EBMUD and Exceltec using sodium sulfite tablets showed less than 10% oxygen depletion.

There are some concerns regarding the integrity of the tablets during dechlorination operations. A field study at EBMUD facility indicated that at high flow rates (475 gpm), sodium sulfite tablets disintegrated, and the remnants escaped the tablet feeder. This lack of dosage control is a potential concern while using this tablet. However, GVRD has been using these tablets for dechlorination under low flow conditions for a few years and has not experienced tablet disintegration. WSSC conducted numerous studies and found that disintegration is a function of the bags used for holding the tablets. When tablets were placed in burlap bags and subjected to full flow fire hydrant flushings the tablets quickly disintegrated. When the tablets were placed in nylon mesh bags, which were less porous than the burlap bags, disintegration of the tablets was significantly reduced.

Sodium Bisulfite ($NaHSO_3$)

Sodium bisulfite is available as a white powder, granule or clear liquid solution. It is highly soluble in water (39%). It undergoes the following reactions with free and combined chlorine:

$$NaHSO_3 + HOCl \rightarrow NaHSO_4 + HCl$$
Sodium Bisulfite + Hypochlorous acid → Sodium bisulfate + Hydrochloric acid

$$NaHSO_3 + NH_2Cl + H_2O \rightarrow NaHSO_4 + NH_3 + HCl$$
Sodium bisulfite + Monochloramine → Sodium bisulfate + Ammonia

$$2\,NaHSO_3 + NHCl_2 + 2\,H_2O \rightarrow 2\,NaHSO_4 + NH_3 + 2\,HCl$$
Sodium bisulfite + Dichloramine → Sodium bisulfate + Ammonia

The reactions produce sodium bisulfate and HCl. NH_3 or NH_4^+ is produced when chloramines are dechlorinated and the speciation depends on pH.

Effect on Water Quality

A one percent solution of sodium sulfite has a pH of 4.3 (Southern Ionics 1998). The production of HCl during chlorine neutralization may marginally decrease pH. Sodium bisulfite is a good oxygen scavenger. Accidental release of slug loads has been reported to have caused injury and death to aquatic species.

Dosage Calculations

On a weight-to-weight basis, approximately 1.45 parts of sodium bisulfite are required to dechlorinate 1 part of chlorine.

Health and Safety Issues

Sodium bisulfite is not carcinogenic or mutagenic and is used in food and drugs as a preservative. The U.S. Food and Drug Administration (USFDA) recognizes sodium bisulfite as safe when used in accordance with good manufacturing practices or feeding practices. However, sodium bisulfite can cause skin, eye and respiratory tract irritation. It is harmful if swallowed or inhaled. Hypersensitivity reactions occur more frequently with asthmatics and bronchial constrictions may also occur. The Occupational Safety and Health Administration (OSHA) Time Weighted Average (TWA) is 5 mg/m^3.

Stability and Reactivity

The strength of sodium bisulfite solutions diminishes somewhat with age. Sodium bisulfite gradually decomposes in air, producing SO_2. It reacts strongly with acids to produce SO_2. Dilution with water also produces SO_2.

Use Calculations

As indicated in the following equation, the feed rate of sodium bisulfite can be calculated, given the flow rate of chlorinated water, chlorine concentration and weight percentage of sodium bisulfite in solution (BetzDearborn 1997).

$$Q_{cfp} = \frac{0.193 * Q_f * Cl}{W}$$

Where,

- Q_{cfp} — Sodium bisulfite feed rate (gpd)
- Q_f — Chlorinated water flow rate (gpm)
- Cl — Chlorine concentration (mg/L)
- W — Weight percent of sodium bisulfite in the feed tank

Cost

The cost of sodium sulfite ranges from $ 0.48 to $ 0.76 per pound (technical grade) if bought in 50-pound containers. Actual costs will depend on quantities ordered and delivery locations.

Other Issues

Sodium bisulfite is available only in crystalline/liquid form. This is less convenient than a tablet form for storage, transportation and handling. However, better control of dosage rates can be obtained by using solutions. Currently, many industries and wastewater utilities use sodium bisulfite solution for dechlorination.

Toxicity to aquatic species caused by sodium bisulfite overdose has been reported by some unidentified wastewater utilities.

Sodium bisulfite may crystallize at room temperatures. It is highly viscous and sometimes difficult to handle. In addition, sodium bisulfite is highly corrosive and caution must be exercised in safely handling this chemical.

Sodium Metabisulfite ($Na_2S_2O_5$)

Sodium metabisulfite is available as crystal, powder or solution. It reacts with chlorine, as well as chloramine, according to the following stoichiometry.

$Na_2S_2O_5$ + 2HOCl + H_2O → $2NaHSO_4$ + 2HCl
Sodium metabisulfite   Hypochlorous acid   Sodium bisulfate   Hydrochloric acid

$Na_2S_2O_5$ + $2 NH_2Cl$ + $3 H_2O$ → $2 NaHSO_4$ + $2 NH_3$ + 2 HCl
Sodium metabisulfite   Monochloramine   Sodium bisulfate   Ammonia   Hydrochloric acid

$Na_2S_2O_5$ + NH_2Cl + $3 H_2O$ → $2 NaHSO_4$ + NH_3 + 2 HCl
Sodium metabisulfite   Dichloramine   Sodium bisulfate   Ammonia   Hydrochloric acid

Reaction Products

Upon reaction with free chlorine in water, sodium metabisulfite produces sodium bisulfate and hydrochloric acid. Dechloramination using sodium metabisulfite produces NH_3 or NH_4^+, depending on pH.

Effect on Water Quality

The pH of 1% solution of sodium metabisulfite is 4.3. Production of HCl during neutralization marginally decreases treated water pH. It is a good oxygen scavenger. Its scavenging properties are comparable to that of sodium bisulfite.

Dosage Calculations

On a weight-to-weight basis, approximately 1.34 parts of sodium metabisulfite are required to remove 1 part of free chlorine.

Health and Safety Issues

Sodium metabisulfite is an eye, throat, skin and lung irritant. Overexposure to sodium metabisulfite can produce highly toxic effects. OSHA Permissible Exposure Limit (PEL) for sodium metabisulfite based on Time Weighted Average (TWA) criteria is 5 mg/m^3. Sodium metabisulfite is poison if entered through intravenous route. MSDS warns of adverse reproductive effects due to over exposure. Ingestion may cause mild to moderately severe oral and esophageal burns. Sodium metabisulfite in food can provoke life-threatening asthma. Hazard ratings are 3,0,1.

Stability and Reactivity

The stability of sodium metabisulfite increases with concentration. It is slowly degraded when exposed to oxygen. Solutions of 2, 10 and 20% strengths are stable for 1, 3 and 4 weeks, respectively.

Sodium metabisulfite reacts strongly with acids to produce SO_2. Dilution with water also produces SO_2. Sodium metabisulfite decomposes at 150° C and produces SO_2.

Cost

The cost of sodium metabisulfite typically ranges from $ 0.48 to $ 0.76 per pound when delivered in 50-pound containers. However, actual cost may vary depending on the volume ordered and delivery location.

Other Issues

Sodium metabisulfite is available only in crystalline/liquid form. This makes it less convenient for storage, transportation and handling than a tablet form-dechlorinating agent. However, better dose control is achieved while using dechlorination solutions.

Its oxygen scavenging properties are a potential concern in field application.

Calcium Thiosulfate (CaS_2O_3)

Calcium thiosulfate is a clear crystalline substance, with little color, a faintly sulfurous odor and near neutral pH. It reacts with free as well as combined chlorine. Calcium thiosulfate undergoes the following reactions with chlorine (Hardison and Hamamoto 1998).

CaS_2O_3 + 2 HOCl + H_2O → $Ca(HSO_3)_2$ + 2HCl
Calcium thiosulfate Hypochlorous acid   Calcium bisulfate   Hydrochloric acid

$Ca(HSO_3)_2$ + 2 HOCl → $CaSO_4$ + H_2SO_4 + 2HCl
Calcium bisulfate Hypochlorous acid   Calcium sulfate   Sulfuric acid   Hydrochloric Acid

In Summary:

CaS_2O_3 + 4 HOCl + H_2O → $CaSO_4$ + 4HCl + H_2SO_4
Calcium thiosulfate Hypochlorous acid   Calcium sulfate   Hydrochloric acid   Sulfuric acid

In addition, the same authors also report a second reaction between calcium thiosulfate and chlorine.

CaS_2O_3 + HOCl → $CaSO_4$ + S + HCl
Calcium thiosulfate Hypochlorous acid   Calcium sulfate   Sulfur   Hydrochloric acid

Although calcium thiosulfate is reported to neutralize combined chlorine effectively, the reactions involved are not currently known.

Impact on Water Quality

A 30% solution of calcium thiosulfate has a pH of approximately 6.5 to 7.5. Chlorine neutralization produces HCl and H_2SO_4 that may result in lower pH. It does not scavenge oxygen and does not produce SO_2.

Aquatic Toxicity

Best Sulfur Products (1997) performed toxicity tests for calcium thiosulfate using fathead minnows. The 96-hour LC50 for fathead minnows is greater than 750 mg/L. Other toxicity information on calcium thiosulfate is not currently available.

Dosage Calculation

Approximately 0.99 mg of calcium thiosulfate is required to neutralize one mg of residual chlorine at pH 7.35. At pH 11, 0.45 mg/L of calcium thiosulfate is sufficient to neutralize 1 mg of chlorine residual. On a weight basis, approximately 1.30 parts are needed per part of chlorine at pH 6.5.

Stability and Reactivity

Calcium thiosulfate is stable under normal conditions. It does not readily release SO_2 gas. However, reaction with acids produces SO_2.

Health and Safety Issues

Hazardous reporting not required. Thermal decomposition may produce SO_2. May cause eye and skin irritation. Hazard rating of calcium thiosulfate is 0,0,0,0 for health, fire, reactivity and persistence.

Cost

Calcium thiosulfate costs about 90 cents a gallon of 30% solution for local supply. Actual cost may vary depending on the quantity and place of delivery.

Other Issues

Calcium thiosulfate does not off-gas SO_2 as other sodium-based dechlorinating agents. It is less toxic to aquatic species. However, dechlorination reactions using stoichiometric concentrations of calcium thiosulfate require nearly five minutes for complete neutralization when it is added (Hardison 1999). Over-dosing of calcium thiosulfate may produce milky-colored suspended solids, causing turbidity violations. Also, excess thiosulfate release may promote thiobacillus bacterial growth. However, bacterial growth is promoted mostly in continuous, excess discharge situations (e.g., cooling water and disinfected wastewater dechlorination operations).

Dechlorination Chemical Summary

Currently, sodium bisulfite, sodium sulfite and sodium thiosulfate are most frequently used by water utilities for dechlorination. The choice of particular dechlorination chemical is dictated by site-specific issues such as the nature of water release, strength of chlorine, volume of water release and, distance from receiving waters. Sodium bisulfite is used by some utilities due to its lower cost and higher rate of dechlorination. Sodium sulfite tablets are chosen by utilities due to ease of storage and handling. Sodium thiosulfate is used for dechlorination since it is less hazardous and consumes less oxygen than sodium bisulfite and sodium sulfite.

However, current knowledge on dechlorination efficiencies of various chemicals is incomplete. A comprehensive study evaluating all the chemicals for various chlorinated water release scenarios is not currently available. Tables 4.1 – 4.3 summarize information currently available on cost, dechlorination efficiency, water quality impacts and regulatory issues. Table 4.1 shows the amount of each dechlorination agent required to neutralize one part of chlorine in distilled water at different pHs. Table 4.2 shows selected regulatory and toxicity information of the dechlorination chemicals. Table 4.3 summarizes relative merits and limitations of the chemicals for dechlorination of water releases.

Sulfur dioxide gas, although widely used in wastewater treatment plants and industrial facilities, is not best suited for water dechlorination for the following reasons. SO_2 is rated as a hazardous and toxic chemical. It is subject to reporting requirements of SARA, Title III, Section 313. Great caution must be exercised in transporting SO_2 cylinders. Also, proper gas delivery and control systems are required for feeding the gas into chlorinated water.

Sodium bisulfite and metabisulfite solutions are also widely used in wastewater treatment plants and industries for dechlorination. Bisulfite solutions are also used as an alternative to hazardous sulfur dioxide for neutralization of chlorinated waters. However, these chemicals are corrosive and they scavenge oxygen from water. An accidental release of these chemicals may cause deleterious effects to aquatic species. Utilities and industries have reported toxic effects to aquatic species due to over dosing or accidental release of bisulfite. These chemicals are highly acidic. The pH of a 1% solution is approximately 4.3. In addition, bisulfite solution crystallizes at temperatures below $44°$ C. This may clog taps and pipes in the chemical delivery system.

Sodium sulfite tablets are used by some utilities for dechlorination of potable waters. It is not rated as a hazardous chemical. One major advantage of this chemical is that it is available in tablet form. Tablets are easier to store and transport. They produce less dust and dissolve more gradually than dechlorination powders and crystals. They are well suited for emergency dechlorination applications. For these reasons, some utilities currently use sodium sulfite tablets for dechlorination. The major limitation in using sodium sulfite is that it may scavenge oxygen in receiving streams. About eight parts of sodium sulfite are required to neutralize one part of oxygen. However, field studies conducted using sodium sulfite at various flow rates indicated less than a 10% oxygen depletion in chlorinated waters. In addition, it is difficult to control the feed rate of tablets. The dissolution rate is not uniform throughout the application. Also, sodium sulfite may off gas sulfur dioxide under heat, which is hazardous.

Sodium and calcium thiosulfate are available in powder/crystalline forms. They are also widely used for dechlorination. The advantages of these chemicals are that they are not rated as hazardous substances. In addition, they are not toxic to aquatic species even at very high doses (LC50 > 750 mg/L for fathead minnows). Tests using sodium thiosulfate showed a LC50 of 850 and 1300 mg/L for ceriodaphnia and daphnia, respectively. Toxicity information for other chemicals and organisms is not currently available. These forms of thiosulfate scavenge less oxygen than the other chemicals discussed. Since they are used in solution form, their feed rate can be better controlled. Calcium thiosulfate is not corrosive and does not off-gas sulfur dioxide. It has the lowest hazard rating of the dechlorinating chemicals used. A limitation of employing these chemicals is that they have to be used in solution form. Hence, the field kit must contain measuring and mixing devices and containers for storage. Flow control using a spigot may not be very accurate. Metering pumps for controlled delivery of solutions are more expensive. The chemical cost of calcium thiosulfate is generally higher than that of sulfur dioxide and other dechlorination chemicals. Over-dosing during receiving stream low-flow periods may sometimes promote thiobacillus and other bacterial growth. However, chances of such bacterial growth are minimal with the intermittent dechlorination applications often encountered in potable water releases.

Table 4.1

Parts of dechlorination chemical required to neutralize one part of free chlorine in distilled water

Chemical	pH 6.6	pH 7.00	pH 9.0
Sodium thiosulfate	2.23	2.13	1.60
Sodium sulfite	1.96	1.96	1.96
Sodium bisulfite	1.61	1.61	1.61
Sodium metabisulfite	1.47	1.47	1.47
Calcium thiosulfate	1.30	1.22	1.08

Based on information obtained from Best Sulfur Products Company, Fresno, CA

Table 4.2

Regulatory information for various dechlorination chemicals

Activity	Sulfur dioxide	Sodium bisulfite	Sodium metabisulfite	Sodium sulfite	Sodium thiosulfate	Calcium thiosulfate
USDOT rating	Poison	Corrosive	Corrosive	Not hazardous	Not hazardous	Not hazardous
NFPA rating*	3,0,0	1,0,1	3,0,1	2,0,0	1,0,0	0,0,0
RMP^c required	Yes	No	No	No	No	No
Exposure limits	NA †	OSHA TWA 5 mg/m^3	OSHA TWA 5 mg/m^3	NA	NA	NA
Aquatic toxicity	NA †	Reported to be toxic	Reported to be toxic	Reported to be toxic	LC$_{50}$$^\pm$ ~ 7.3 g/L	LC$_{50}$§ > 750 mg/L

* National Fire Protection Authority Rating for health, fire and reactivity, respectively. (Ranges from 0 to 4. 0 - No hazard, 4 - extremely hazardous)
† Not available
$^\pm$ USEPA (1988)
§ 96 hour test for fathead minnows

Information in Table 4.2 is modified from various material safety data sheet(s) for chemicals.

Table 4.3

Comparison of various agents for dechlorination of potable water releases

Activity	Sulfur dioxide	Sodium bisulfite	Sodium metabisulfite	Sodium sulfite	Sodium thiosulfate	Calcium thiosulfate
Form available	Gas	Powder/crystal	Powder/crystal	Powder/crystal and tablet	Powder/crystal	Powder/crystal
Cost ($/lb) *,†	1.37 ‡	0.48	0.48	0.50 §	0.70	1.06
Dose at pH 7.0 (mg/mg Cl) *,††	.99	1.61	1.47	1.96	2.23**	1.43**
Cost($)/MG/ ppm Cl*	11.40	6.42	5.86	8.15	12.98	12.60
pH of 1% solution	NA ‡‡	4.3	4.3	8.5-10.0	7.0	6.5-7.0
Decomposition/off-gassing	SO_2	SO_2	SO_2	SO_2	SO_2	No SO_2
O_2 scavenger (mg/mg O_2)	Strong	Strong	Strong	Moderate	Weak	Weak
Crystallization	N/A §§	Yes	Yes	No	No	No

* Based on information from Skagit County, WA.

† 1999 cost in U.S. dollar. Actual cost may vary with quantity and delivery location.

‡ Cost corresponds to 50 lb cylinder. Cost decreases significantly for larger tanks.

§ Cost of sodium sulfite tablet is ~ $ 2.75/lb (Source: WSSC).

** Dose decreases with increase in pH.

†† After 10% addition

‡‡ Not Available

§§ Not applicable

CHEMICAL FEED TECHNIQUES

In this section, various methods currently used to feed dechlorinating agents into chlorinated water are described.

Gravity Feed Method

The gravity feed method typically involves adding dechlorinating solution from a container equipped with a spigot that would be placed on the curb above the water flow path. The discharge spigot on the container can be adjusted to provide a minimum dechlorinating solution feed rate into the water flow, based on calculations involving the concentration of the dechlorinating solution, water flow rate, and residual chlorine concentration in the flow stream. In order to minimize the volumes of dechlorinating solution needed on field vehicles, wherever possible, dry dechlorinating agent should be mixed directly within the container prior to use, rather than using pre-mixed dechlorinating solutions provided by suppliers. As part of the dechlorination procedure, samples can be collected downstream of the feed point, analyzed for pH, dissolved oxygen and residual chlorine and, if necessary, the chemical feed rate can be adjusted to ensure a non-detectable residual chlorine concentration prior to discharge to the surface water.

Gravity feed systems are simple to operate, have minimal equipment requirements, have been used effectively by various utilities, and are inexpensive (low-density polyethylene carboys equipped with spigots and having a capacity of 5.5 gallons are currently available for about $80). Unless adjusted, chemical feed rates can be expected to decrease slightly over time, however, as the available head pressure within the carboy decreases during use. A disadvantage of using this technique is that, it involves field testing and calculations for flow rates and water quality parameters to adjust chemical feed rate. Field maintenance crews sometimes prefer a method that does not involve field calculations.

Chemical Metering Pumps

This method for injecting a dechlorinating solution is similar to the above-described gravity feed method, except that a chemical metering pump is used to inject the dechlorinating solution from a container into the water flow. Chemical metering pumps are capable of delivering chemical solutions over a wide range of flows (e.g., 0.0006 to 3,400 ml/minute), the flow rates are adjustable, and the pumps provide a constant chemical feed rate.

Relative to a gravity feed system, this type of feed system is more difficult to operate, requires more equipment (e.g., storage container, pump, energy source, and tubing) and costs are significantly higher. Although chemical-metering pumps may provide field personnel with a more reproducible method, this feature involves much higher cost and greater operator ability and attention during dechlorination. The cost of metering pumps is approximately $750 per pump for a 115 volt AC model and about $850 per pump for a 12 volt DC model (including pump head, drive, and tubing). In addition, the tubing requires periodic replacement over time.

Venturi Injector Systems

Venturi injectors are differential pressure injection devices that allow for the injection of liquids (e.g. dechlorinating solutions from hydrants, etc.) into a pressurized water stream. This method is well suited for pressurized water releases such as those through hydrants. Pressurized water entering the injector inlet is constricted toward the injection chamber. This results in a higher velocity water stream through the injection chamber than at the injector inlet. The increase in velocity through the injection chamber results in a decrease in pressure, thereby allowing the dechlorinating agent to be drawn from a storage container through the suction port and entrained into the water stream.

Water main discharges would be constricted by a regulator valve or gate valve and routed through a Venturi injector system. Dechlorinating solution would be drawn into the injector from a plastic container, with the chemical feed rate controlled by a metering valve. In addition to the valves shown on the schematic diagram, Venturi injector units should include a flow meter installed near the metering valve to measure the chemical feed rate (e.g. a rotameter), a threaded fitting at the upstream end of the pipe to attach adapters (e.g. reducing sections), and a fitting at the downstream end of the pipe for attachment of flexible discharge piping (e.g. a hose). Arden Industries (Shingle Springs, CA) supplies Venturi injector systems for hydrants at a unit cost of about $1,090 and for main discharges at a unit cost of about $750.

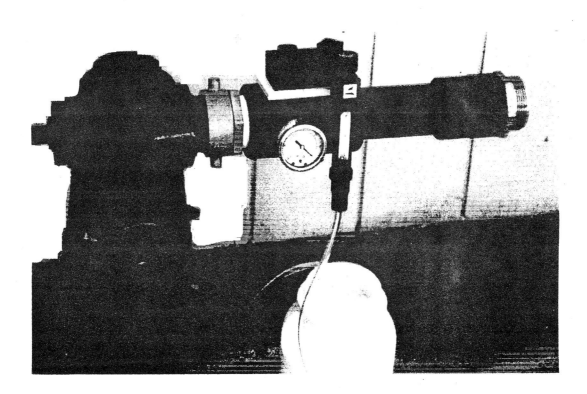

Source: Arden Industries, Shingle Springs, CA

Figure 4.1 'Bazooka' Venturi dechlorination feeder by Arden Industries (patent pending)

The primary advantages of Venturi injection systems are:

- efficient operation over a wide range of pressures;
- available in a wide range of sizes, flows, and injection capacities;
- no external energy requirements; and
- instantaneous mixing via cavitation in the injection chamber.

The primary disadvantages of Venturi injection systems are:

- more sophisticated equipment requirements than gravity feed systems;
- slightly more labor intensive set-up;

- higher unit equipment cost than gravity feed systems; and
- may require constant monitoring.

Spray Feed Systems

The dechlorinating solution can be sprayed into the flow via a backpack sprayer similar to those used for pesticide and herbicide application. The advantages of this technique are that the chemical feed rate is fairly constant (given a steady pressure within the solution chamber). Hence, dosages can be approximated fairly accurately, and a piped or channelized flow is not required to effectively feed the chemical. This method is typically more effective in adding dechlorinating chemicals to sheet flows than the other alternatives evaluated above.

A significant disadvantage of a spray feed system is that it requires the equipment to be set up at a stationary point and monitored for adequate chemical and pressure. If a stationary point is not available, a person will have to don the sprayer and continuously apply the dechlorinating agent to the flow, which is very labor intensive.

Flow-through Systems

Flow-through systems include any method where the solid chemical is held stationary and the flow allowed to run over, around and/or through it. Examples include pumping chlorinated water through a container filled with dechlorinating agent, or laying permeable bags of the chemical in the flow path. For this application, it is anticipated that the dechlorinating agent would be used in tablet or powder form. The advantages of flow through systems are that they are simple and can be used for sheet flow applications as well as for channelized or pumped flow. The disadvantages are that there is no control over the dosage, overdosing or underdosing to significant levels could easily occur and it may be difficult, in some cases, to tell when the chemical has been used up and must be replaced. Also, due to the contact time required for dissolution of powder/tablet, this method is more suitable for low and medium velocity discharges. This method may not be suitable for dechlorinating releases from unidirectional flushing where the velocity of the flow is approximately 5.0 feet/second. Some variations in the application of flow through systems are described below:

Automatic Tablet Dispensers

Automatic tablet feeders for dechlorination are similar to the tablet feeders currently used for disinfection in many water/wastewater treatment plants. The feeders typically consist of a non-moving housing and automatic feed tubes. The tubes are inserted down through a removable top cover of the feeder into the stream of water. The lower end of each tube is slotted to permit free flow of water through the tubes to assure good contact between the water and dechlorination tablets. The feeders typically contain a removable wire plate at the outlet end to control the internal water level based on the flow rate and level of dechlorination required. Proper connection is required to direct the flow to the feeders, which typically have 6-inch pipe inlet or solid inlet end for field adaptation.

As the stream of water flows past the feed tubes containing dechlorination tablets, the dechlorination agent is released into the water by dissolution. At the outlet end, a weir controls the height of the water level in the feeder, which controls the concentration of the agent in the water, regardless of surges in the water flow entering the feeder. As the incoming water flow rate increases, the water level in the unit rises, immersing a greater number of tablets. Since the amount of agent dissolved depends upon the number of tablets immersed in the water, the dechlorination agent level remains constant regardless of the water level in the feeder.

The feeders commonly used by utilities and industrial facilities that provided information for this report are built with medium density polyethylene, with a dimension 26 inches long, 18 inches wide and 16 inches deep. Each feeder is equipped with up to four feeder tubes, 24 inches long with an outside diameter of 3.5 inches. The feeders can handle up to an average flow rate of 35 gpm and a peak flow (4 hour) of 87 gpm. Hence, they are better suited for dechlorination of smaller chlorinated water discharges. Multiple feeders can be used for dechlorinating higher flows. Currently, Exeltec and Norweco companies manufacture automatic tablet feeders. Both these companies make their brand of sodium sulfite tablets. The addresses of the companies manufacturing the feeders are included in Appendix A.

Other Flow-through Systems

Washington Suburban Sanitation Commission has developed and patented several flow through devices suitable for use under different flow conditions. In addition, EBMUD has developed several flow-through devices suitable for dechlorinating a wide range of flows from various sources, using sodium sulfite tablets. The details are provided in Chapter 5 (New Technologies).

FLOW CONTROL MEASURES

During planned and unplanned water releases, it may be necessary to construct flow control measures to prevent the water from entering directly into a water body and to provide an opportunity for better mixing of the dechlorinating agent. GVRD recommends construction of berms, swales, ditches or redirection pipes to control the flow of released water.

Berms

Berms can be constructed using sand bags, hay bales, gravel with a filter fabric core, plywood or similar materials. Sand bags are often used by utilities to construct temporary berms. Sand bags placed in a semicircle, with a depression in the ground upstream, may provide a pond-like structure with a higher residence time for dechlorination. Sandbags can be made from material available at site or that brought to site.

Swale

A swale or natural depression near the point of chlorinated water release can be used to control the flow of water. If such a setting is available near the flow, the water can be redirected towards it. Constructing a berm at one end of the swale could significantly increase the holding capacity.

Ditches

Management practices developed by GVRD involve directing chlorinated water flow to ditches nearby to allow for better mixing and dechlorination. If available, ditches near chlorinated water release locations can be used as temporary holding ponds. Check dams can be constructed using sand bags or hay bales to improve mixing. The flow of water can be directed toward the ditches using sand bags and hay bales.

However, caution must be exercised in using this technique since, in many places, ditches may be considered to be waters of the state for which WQS apply. Clearly, this approach would not be acceptable in such a situation.

Redirection Pipe

A redirection pipe can be used to redirect the flow of water to a specific area or into a holding structure. It could also be used as a treatment system if an in-line injection system is used to inject the dechlorinating agent into the pipe.

GVRD recommends polyvinyl chloride (PVC) pipe for constructing redirection pipes. These pipes are not expensive and are readily available in different diameters. If the in-line injection method is used, there should be an injection point at the end closest to the source of the chlorinated water. This permits the dechlorinating agent to mix within the redirection pipe prior to entering the receiving water.

The injection point can be a vertical pipe that intercepts the discharge pipe. It can be constructed by providing a T-joint into the pipe. The vertical pipe should be short enough (less than 1 m) to minimize lifting of chemicals by the operators during the process. A chemical metering pump can be fitted to the top of the injection point to control chemical dose rate.

FIELD METHODS FOR RESIDUAL CHLORINE MEASUREMENT

In this section, several alternative methods for measuring residual chlorine in the field are described. Monitoring residual chlorine concentrations is a requirement during most dechlorination efforts. The concentration of residual chlorine will indicate if the discharge requires dechlorination, and whether the dechlorination operation is effective.

Most field methods for measuring chlorine concentrations use an indicator that alters the color of the water sample. Upon addition of the indicator, the color of the water sample is compared to a set of colored standards. For example, solutions containing higher concentrations of chlorine turn darker than those containing lower chlorine concentrations.

The primary evaluation criteria in selecting an appropriate field method include ease of use, detection limits, precision, accuracy, and cost. The selected method should provide data of acceptable quality for documenting that the Field Management Practices adopted are effective in meeting regulatory discharge limits. The limitation in most of the field kits is the detection limit. In most cases, the detection limit of the test kit is about 0.1 mg/L, which is above the regulatory discharge level for many jurisdictions. Free chlorine concentrations of 0.03 mg/L to 0.05 mg/L have been reported to be toxic to many aquatic species. Although these kits are not accurate below a chlorine concentration of 0.1 mg/L, upon addition of the indicator the samples turn faintly colored at much lower chlorine concentrations. This slight change in color is used as an indicator of the presence of trace concentrations of chlorine.

Some field methods for measuring residual chlorine concentrations are described below. The costs of equipment and reagents provided are 1999 costs.

Water Quality Test Strips

Water quality test strips are available for measuring free chlorine and total chlorine. Typical test ranges and costs for test strips are as follows:

Test	Test Range	Cost/50 Test Strips
Free Chlorine	0.5 to 2.0 mg/L	$ 11.00
Total Chlorine	0.5 to 5.0 mg/L	$ 11.00

Test strips are easy to use, provide results within 90 seconds, and require no reagents. Analyses are performed by immersing the test strips in water samples and comparing the resultant colors visually against color standards. These strips can measure down to 0.5 mg/L residual chlorine. Although easy to use, test strips are less accurate and reliable for most dechlorination requirements, and are not generally recommended for this application.

Swimming Pool Test Kits

Swimming pool test kits are readily available for measurement of residual chlorine by visual colorimetric methods at a relatively low cost (approximately $12). Free, combined, and total chlorine are typically measured by adding N,N-diethyl-p-phenylenediamine (DPD) indicator.

This method typically involves filling sample test tubes to graduated levels, adding a fixed number of reagent drops, mixing, and visually matching the resultant colors against color standards that are included in the test kits. This method is not sensitive for analysis below 0.5 mg/L. It is less accurate than other field methods described and the detection limit using this method is high (0.5 mg/L). Hence, this method may not be suitable for measuring chlorine concentrations for dechlorination activities. However, such test kits have potential applicability in identifying the presence or absence of chlorine residuals, and in measuring approximate concentrations of free, combined, and total chlorine.

Orthotolidine Indicator Kit

One available field sampling kits uses a liquid orthotolidine indicator for residual chlorine analysis. The kit contains sample vials of preset volumes. The water to be tested is transferred into the sampling vial and then a pre-measured amount of indicator is added. The color of the water upon the addition of the indicator is compared with that of a sheet of paper supplied with the kit or with a color disk to determine the residual chlorine concentration.

The detection limit of this kit is about 0.1 mg/L. The cost of the kit is about $110. However, GVRD has expressed some concerns over health and safety issues (potential carcinogen) related to the chemicals used in this test. In addition, Hach's (1995) manual refers to some studies which indicated that the orthotolidine method gave poor accuracy and precision and high overall error in comparison with other chlorine methods. This method was dropped from the 14th Edition of Standard Methods for Waster and Wastewater Analysis. Hence, this method is not recommended for field chlorine analyses.

Field Colorimetric Test Kits

Field test kits are available for the measurement of a broad range of water quality parameters, including residual chlorine. Several equipment companies supply Pocket/portable colorimeters capable of measuring free and/or combined chlorine. The range of residual chlorine that can be measured varies with the manufacturer and the model of the colorimeter. Typically the range varies from 0 to 2 mg/L to 0 to 4.5 mg/L. Pocket colorimeters are considered accurate for measuring chlorine concentrations greater than 0.1 mg/L. In this technique, a pre-measured amount of DPD reagent is added to the water sample, mixed well, and the sample analyzed for chlorine concentrations using a portable colorimeter. The liquid crystal detector indicates the chlorine concentration in solution.

An incomplete list of equipment vendors includes Hach (Loveland, CO), Hanna Instruments (Newfoundland, NJ) and Orion (Beverly, MA). Four AAA alkaline batteries with a typical lifetime of 750 tests power the Hach field test kits. The reported accuracy at 25 $^\circ$C is +/- 0.02 mg/L. The cost of the colorimeter is approximately $299. Twenty-five individually wrapped packages of DPD reagents for free or total chlorine measurements are available at a cost of $13.60. The Hanna Instrument colorimeters are powered by a 9-volt battery, with a typical lifetime of 300 tests. The reported accuracy between 0 to 50 $^\circ$C is +/- 0.01 mg/L. A container with a hundred individually wrapped packages of DPD reagents for free or total chlorine measurements is available at a cost of approximately $14.00.

It is anticipated that such test kits would be capable of producing technically defensible residual chlorine measurements over the range of concentrations likely to be encountered during potable water discharges. Hence, this method is widely practiced and recommended for residual chlorine analyses during potable water releases. However, measurement of residual chlorine concentrations during disinfection of new and repaired mains (which may be as high as 200 mg/L) would require sample dilution prior to analysis.

As an alternative to using a colorimeter, some water utilities add DPD pillows to water samples and rely upon visual observations of the lack of color development to indicate that dechlorination has been effective. However, this approach can potentially result in overdosing of the dechlorination agent.

DPD Titration Method

This is a titration method that involves the reaction of chlorine with DPD (APHA, AWWA, WPCF 1989) (4500 Cl F. DPD Ferrous Titrimetric Method). Upon DPD reaction, the colored sample is titrated with a ferrous reducing agent to the colorless endpoint. A digital titrator is used for titration. The number of digits required to the end point is used to calculate the chlorine concentration.

This method may be slightly advantageous in measuring chlorine concentration in colored wastewaters. However, this procedure requires additional time for analysis. Accurate measurement of sample volume for titration is essential. To achieve accuracy, a pipette may be used, a procedure which can lead to loss of volatile chlorine species. The visual estimates of the titration endpoint may be less precise compared to the measurement of color obtained by using a colorimeter or spectrophotometer. Hach Company reports an analytical range of 1-3 mg/L for this method with a detection limit of 0.02 mg/L in laboratory samples.

Amperometric Titration Method

Standard Methods for the Examination of Water and Wastewater (1989) (4500 Cl D. Amperometric Titration Method) recommends amperometric titration methods for accurate analysis of free and combined residual chlorine concentrations. A significant advantage of this method is a very low detection limit (0.01 to 0.02 mg/L) and the ability to measure chlorine in waters containing high turbidities (e.g., discharges during trench dewatering). This method, however, is not as simple as the above-described colorimetric methods and requires greater operator skill. This method is generally performed in a laboratory.

The amperometric titration method involves using an amperometer to measure changes in conductance of the sample during titration. When titrant is added, halogens (chlorine) in the sample are reduced and the sample conductance decreases. The titration endpoint is reached when the current stops decreasing with further addition of titrant. Titration for free chlorine is performed at a pH of 6.5 to 7.5, while titration for total chlorine is performed in the pH range of 3.5 to 4.5, after adding potassium iodide to the sample. Detection limits for free and total chlorine residuals are 0.01 mg/L and 0.02 mg/L, respectively.

Compared to the methods described earlier for measuring chlorine, the equipment and reagents required for the amperometric method are expensive and relatively complex. The equipment required for field measurements of total and free chlorine includes a 250 ml graduated cylinder, a portable amperometric titrator (e.g., Fischer & Porter Series 17T2000), and a 250 ml beaker. Required chemical reagents include standard phenylarsine oxide titrant (0.00564N), phosphate buffer solution (pH 7), potassium iodide solution, and acetate buffer solution (pH 4). The cost of a fully equipped Fischer & Porter Series 17T2000 portable amperometric titrator capable of performing 120 titrations is currently about $1,500.

The United States Enrichment Corporation (Patuka, Kentucky) uses the amperometric method for field chlorine measurements. However, routine use of the amperometric titration method for chlorine measurements during water main discharges is not recommended for the following reasons:

- The amperometric method is significantly more difficult to perform than colorimetric methods.
- The method requires greater operator skill to obtain reliable results.
- Loss of chlorine can occur because of rapid stirring.
- The equipment and labor costs associated with performing this test are significantly greater than for colorimetric methods.
- The preparation of chemical reagents required to perform this test is more difficult than for other colorometric methods.
- The increased precision and accuracy that this method provides is often not necessary for the routine evaluation of the effectiveness of dechlorination activities.

- The equipment is delicate and, although a field model exists, it is not suitable for continuous use under the range of conditions encountered during dechlorination of the various types of discharges.
- It is recommended that personnel untrained in laboratory procedures not use the equipment.

Analytical Interference

The field methods for chlorine analysis may give false readings in some circumstances. For example, tetravalent manganese is reported to interfere with the DPD indicator method and provide a false reading. Turbidity and waters with initial colors may also interfere with chlorine analysis by this technique. Reports indicate that the presence of sulfite ion, used for dechlorination, may interfere with analysis of chlorine using colorimetric techniques (Helz and Nweke 1995). Furthermore, molybdenum is reported to interfere with chlorine analysis by colorimetric techniques.

Hach Company (1995) has published a technical information series booklet on chlorine analyses and interference. Hach indicates that in most cases concentration of these ions and suspended particles in potable waters may not interfere with chlorine analysis. However, in the case of manganese interference, the Hach Company recommends an initial dechlorination of samples with sodium arsenite, which does not affect manganese, before proceeding with the test. The result obtained with the dechlorinated sample is subtracted from the normal test result to obtain the correct chlorine concentration. Caution must be exercised, however, in safe disposal of arsenite-containing test waters, if this method is used to minimize manganese interference.

To minimize interference due to turbidity, Hach recommends filtering the samples after reaction with DPD is completed. The color (Wurster dye) is positively charged. Hence, the filter must be inert to avoid sorption of color to the material. Hach reported satisfactory results using a 3-micron filter for turbid waters. The procedure involves filtering a portion of the sample for calibration (zeroing of the photometer). The test sample is treated with DPD, filtered upon completion of reaction (color formation), and analyzed for chlorine.

If the water is highly colored, Hach recommends the DPD titration method or amperometric titration method for chlorine analysis.

Whenever possible, chlorine free samples of the waters to be tested must be analyzed to establish a false reading. This can be achieved by agitating the target sample to reduce its chlorine concentration prior to analysis. If the color of the sample still persists, it can be repeatedly agitated to ensure chlorine removal. If positive reading still persists, it is likely that interference is present and must be accounted for during chlorine monitoring.

In addition, concentrations of chlorine above 5 or 6 mg/L can lead to the formation of a colorless imine product with the DPD reagent, rendering colorimetric methods ineffective. Hach (1995) recommends dilution of samples prior to analysis with DPD or analysis using DPD/amperometric titration techniques at higher chlorine (> 4 mg/L) concentrations.

Summary

Table 4.4 summarizes the benefits and limitations of various chlorine measurement techniques. The water quality strips and swimming pool kits are easier to use and less expensive than most other techniques. However, they are not sensitive for measurement below a chlorine concentration of 0.5 mg/L. Hence, these techniques are not recommended for dechlorination activities. Orthotolidine method is reported to be sensitive for measurement down to 0.1 mg/L residual chlorine. However, some health & safety concerns exist regarding the use this chemical. Pocket colorimeters using DPD can measure a residual chlorine concentration as low as 0.1 mg/L accurately. This technique can analyze free and/or total residual chlorine concentrations. It is easier to use and samples can analyzed within a few minutes. This technique is widely used by utilities for residual chlorine analysis. However, colored and turbid samples may cause some interference during analysis. The Hach Company recommends filtering the samples after the reaction with DPD to minimize interference due to turbidity using this technique.

DPD titration and amperometric titration techniques have a lower detection limit (0.02 mg/L) for residual chlorine measurement. However, both of these techniques require more time and greater operator skill. DPD titration technique involves reaction of the sample with DPD followed by titration with a ferrous agent to a colorless endpoint. To achieve accuracy, sampling with a pipette may be required. However, this procedure may result in loss of volatile chlorine species. Both these techniques are better suited for analysis of residual chlorine in the laboratory.

Table 4.4

Comparison of chlorine measurement techniques

Technique	Platform	Test range (mg/L)*	Detection limit (mg/L)	Comments
Water quality strips	Visual comparison with standards	0 – 2.0 † 0 – 5.0 ‡	0.5	Not suited for dechlorination due to higher detection limit
Swimming pool kit	Visual comparison with standards	0 – 2.5	0.5	Not suited for dechlorination due to higher detection limit
Orthotolidine technique	Visual comparison with standards	0 – 2.0	0.1	Some health & safety concerns exist
Pocket colorimeter	Colorimeter with liquid crystal detector	0 – 2.0	0.1	Potential interference due to color and turbidity
DPD titration	Digital titrator to colorless endpoint	0 – 3.0	0.02	Requires more time and greater operator skill
Amperometric titration	Conductance measurement	0 – 3.0	0.02	Requires more time and greater operator skill

Source: Hach Company (1995) *Technical Information Series – Booklet No. 17*, and Kennedy/Jenks Consultants
* - For free and total residual chlorine, unless mentioned otherwise
† – Free chlorine
‡ – Total chlorine

DECHLORINATION INFORMATION IN AWWA STANDARDS

Currently, AWWA does not have a standard for dechlorination of potable waters. AWWA has provided the following information for disposal of heavily chlorinated water in Appendix C of the Standard for Disinfecting Water Mains (ANSI/AWWA C651-99). This is provided by AWWA as information only and not as part of the AWWA C651.

1. Check with the local sewer department for conditions of disposal to sanitary sewer.
2. Chlorine residual of water being disposed will be neutralized by treating with one of the chemicals listed in the Table 4.5.

Table 4.5

Amounts of chemicals required to neutralize various residual chlorine concentrations in 100,000 gal (378.5 M^3) of water

Residual chlorine conc. (mg/L)	Sulfur dioxide (SO$_2$)		Sodium bisulfite (NaHSO$_3$)		Sodium sulfite (Na$_2$SO$_3$)		Sodium thiosulfate (Na$_2$S$_2$O$_3$ 5H$_2$O)	
	(lb)	(kg)	(lb)	(kg)	(lb)	(kg)	(lb)	(kg)
1	0.8	0.36	1.2	0.54	1.4	0.64	1.2	0.54
2	1.7	0.77	2.5	1.13	2.9	1.32	2.4	1.09
10	8.3	3.76	12.5	5.67	14.6	6.62	12.0	5.44
50	41.7	18.91	62.6	28.39	73.0	33.11	60.0	27.22

Reference: Table C.3, *AWWA Standard for Disinfecting Water Mains*, C651-99, published by the American Water Works Association. 1999.

CURRENT DECHLORINATION PRACTICES BY UTILITIES

In this section, dechlorination practices of water utilities participating in this project are summarized. Practices of some other utilities that are not participating in this project are also included. In addition, details of cooling water discharge dechlorination procedures followed by two industrial facilities in Tennessee and Kentucky are described. These are the methods practiced by the utilities/industrial facilities at the time of preparing this report. Note that utilities continually change and update their dechlorination practices in response to regulatory and other imperatives. The reader should contact the agency for the most current practices. The addresses of agencies whose methodologies are described in this report are provided in Appendix A.

U.S. Utility Practicing Extensive Dechlorination

Among the U.S. utilities participating in this project, East Bay Municipal Utility District, City of Portland, Bureau of Water Works and Washington Suburban Sanitation Commission dechlorinate all planned releases.

East Bay Municipal Utility District (California)

Dechlorination Management Practices followed by EBMUD are summarized below.

Waters requiring dechlorination. All discharges from distribution systems except those listed below as exempt discharges are dechlorinated. This includes trench-dewatering water when containing leakage from water mains, mains or hydrant flushing water, hydrant testing water, aqueduct dewatering and other planned releases.

Exempt discharges. Discharge of raw, untreated water with no chlorine residual, discharges of groundwater, discharges from activities producing less than 1000 gallons, unplanned discharges from unmanned locations, discharges to land, leakage and on-hold repairs not under the District's control are not dechlorinated.

Dechlorination agents. Following conversion to the use of chloramines as a disinfectant, EBMUD initially used sodium thiosulfate for dechlorination. Sodium bisulfite solutions were also used, and continue to be used, for dechlorination activities by certain District work groups. Although EBMUD has subsequently adopted the use of sodium sulfite tablets for many dechlorination applications, as discussed below, sodium thiosulfate and sodium bisulfite solutions continue to be used for certain dechlorination applications. An 11.5% and a 23% solution are used individually or in combination based on flow rate and volume.

Priority of operation. While dechlorinating using thiosulfate, the highest priority is given to worker health and safety, then to public health and safety, and finally to environmental protection.

Dechlorination using sodium thiosulfate. Dechlorination is accomplished by the addition of sodium thiosulfate solution to the discharge flow. A solution is made by mixing a fixed, pre-packaged amount of crystalline sodium thiosulfate with water in a polycarbonate container equipped with an adjustable spigot (called a carboy). The carboy is placed above the flow path and the solution fed into the flow by gravity. This is done by setting the carboy on the curb or other location out of the way of field crew operations and as far upstream of the storm drain inlet or receiving water (whichever the flow enters first) as possible. This is done so that the chemical solution drains from the carboy into the flow path at a point where good mixing with the entire flow stream will occur (i.e., where the flow is at its deepest and narrowest).

If the discharge is segregating into two or more flow paths, sandbags are used to direct flow into one distinct channel. If unifying the flow is not possible, and the branch streams represent a significant flow (>5 gpm), one or more (depending on the number of flow paths) additional carboys are positioned to dechlorinate the additional streams. Solution concentrations and feed-rates are varied according to the rate, duration and chlorine concentration of the discharge.

Dechlorination using sodium sulfite. Dechlorination of chlorinated water discharges is accomplished by the addition of tablets comprised of 90% sodium sulfite to the discharge flow. For discharges from trenches during main breaks, the tablets are placed inside synthetic mesh fabric pockets sewn together in a grid or line (called a "dechlor mat" or "dechlor strip" respectively; Figures 5.1 [p.125] & 5.2 [p.126]). The dechlor mat or strip is laid across the flow path or over the storm drain and either weighted down or nailed to the street to keep it in place. For discharges from hydrants or blowoffs, tablets are either placed inside a chamber on the flow diffuser or inside synthetic mesh pockets attached to the discharge face of the diffuser (Figure 5.3 [p.127]). In all cases, as the discharged water flows over and around the tablets, chemical is released as the water contacts the tablets, reacting with and destroying the chloramines. The key to the success of this procedure requires effective contact between the flow and the tablets. This is accomplished by ensuring that the tablets are well distributed across the flow path. Testing indicates that the tablets should be spaced no more than 4" apart for gravity discharges at ambient pressure. For discharges under pressure, the tablets should be spaced as close together as possible without constricting the flow. The various tablet holder designs are fabricated to ensure that this specification is met.

The decision to use the mat or strip is made on a case-by-case basis. Both choices have benefits and limitations. The mats can cover a larger area, so if the discharge flow is large and spread out, mats may be easier to use than multiple strips. Mats are also sized to cover storm drain inlets so if the flow is not well channelized, it may be easier to locate mats over the storm drain(s) that receive the flow, rather than laying out strips or mats upstream of this point. Strips are smaller, take up less space in vehicles and multiple strips can be used to cover larger flows so their convenience and flexibility make them the appropriate choice unless some of the conditions described above are encountered.

Equipment needed for dechlorination includes dechlor mat (3' x 4') or dechlor strip (6" x 36") or Diffuser with tablet chamber or diffuser with mesh tablet holder; dechlor tablets (45 lb bucket); DPD Powder-Pop Dispenser; and personal protective equipment (goggles and rubber gloves) (See Chapter 6 for details).

Monitoring dechlorination efficiency. Total chlorine is measured colorimetrically by adding DPD reagent to a sample of a given discharge (Pocket colorimeter). If a pink or red color develops when the reagent is added to the sample, chlorine is present. Absence of color indicates there is no detectable chlorine present.

Washington Suburban Sanitary Commission (Maryland)

Washington Suburban Sanitary Commission *(WSSC)* at Maryland is another utility practicing extensive dechlorination. WSSC has developed several standard operational procedures as well as best management practices (BMPs) for disposal of chlorinated water. In addition, WSSC has developed patented dechlorinating techniques/devices for dechlorination.

Waters requiring dechlorination. WSSC dechlorinates all planned water releases except flow test and meter test waters. Unplanned water releases are not currently dechlorinated.

Dechlorination agent. WSSC predominantly uses sodium sulfite tablets for dechlorination.

Dechlorination methods. For any type of chlorinated water release, discharge to sanitary sewer is the preferred mode of disposal for WSSC. However, if for any reason it is not possible, chemical-dechlorination procedures are followed.

WSSC primarily uses dechlorination techniques developed in-house (tablet/suspension, tablet/diffuser, tablet/plastic pipe and Tablet/Cage methods). These techniques are currently being patented. Details regarding these techniques are presented in Section 5 (New Technologies).

Dechlorination of super-chlorinated waters. WSSC has special requirements for dechlorination of super-chlorinated waters. Super-chlorinated waters are those waters containing residual chlorine concentrations of greater than 5 mg/L. Generally, activities such as disinfection of repaired mains as per AWWA disinfection standards produce such waters.

The preferred method of disposal of such waters is discharge into sanitary sewers. However, WSSC has established a flow rate limit of 0.25 MGD (million gallons per day) for discharging such waters into sanitary sewers to prevent concentrated slugs of chlorinated water from reaching WSSC's wastewater treatment plants. This rate may be increased if necessary through coordination with WSSC Wastewater Operations Division.

WSSC sometimes uses existing sediment control ponds for dechlorination.

If neither a sanitary sewer nor a holding pond facility is available, WSSC requires the water be handed over to a contractor for off-site dechlorination.

City of Portland, Bureau of Water Works (Oregon)

The City of Portland dechlorinates selected releases in accordance with the guidance provided in the Department of Environmental Quality - Management Practices for Disposal of Chlorinated Water.

Target waters. Dechlorination is practiced on all planned releases when not discharged into sanitary or combined sewer systems. When there is enough time to react, unplanned releases are also dechlorinated.

Exempted waters. Chlorinated waters that are discharged to sanitary or combined sewers and those with at least 1000 feet of travel before discharging into receiving streams are unregulated for chlorine residual. New main disinfection waters and waters from rechlorination of repaired mains contain high chlorine concentrations (> 4 mg/L). In most cases these waters are discharged to sewer lines. Unplanned waters with insufficient time to react (fire fighting and water main breaks) are not dechlorinated. Discharges less than 500 gallons per event are exempted. According to the ODEQ dechlorination guidance, releases to receiving waters with more than 50 cubic feet per second flow will probably not require dechlorination.

Although these waters are currently not regulated for chlorine residual in Oregon, it is the responsibility of the discharger to ensure that the chlorine concentration in all released waters is below 0.1 mg/L before discharging into receiving streams.

Dechlorination goals. As per ODEQ guidelines, chlorine concentrations in the discharge water should be less than 0.1 mg/L.

Dechlorination agents. Sodium sulfite tablets and granulated sodium thiosulfate are used.

Dechlorination practices. The most common dechlorination method is to place a number of "D-Chlor" tablets (sodium sulfite) in a burlap bag and place the bag in the middle of the flow some distance downstream of the discharge point. More bags are used for larger flows. There is no set dosage and more tablets or bags are added according to total chlorine readings taken downstream. This procedure is used for flushing, supply conduit draining, hydrant flow testing, flushing associated with main disinfection, and some instances of tank "freshening".

There are several "D-Chlor" tablet feeder installations at locations where water is being drained constantly at low flows for "freshening". The tablets are replenished weekly and the residual is taken to ensure that adequate dechlorination is being achieved.

There are 72 tanks and standpipes in the City of Portland distribution system. Technical grade anhydrous sodium thiosulfate in liquid or granulated form is used for de-chlorinating the contents of a tank prior to draining. Use of the liquid or granules depends on tank characteristics. Liquid sodium thiosulfate is applied at a ratio of 1 part to 40 parts water. Granules are applied at a rate of five pounds sodium thiosulfate to one pound chlorine.

Disinfection of new mains or rechlorination after main repairs involves the disposal of super-chlorinated water. ODEQ defines this as water having a chlorine residual greater than 4 mg/L. Super-chlorinated water is discharged to the sanitary sewer whenever possible. If that option is not available, then the water is treated by other means. In some cases water is pumped to a water tanker and a dechlorinating agent is added to the tank. That water is tested and if it meets the residual chlorine standard of 0.1 mg/L or less, it is discharged to waste or used for construction purposes or street cleaning. After disposing the super-chlorinated water, the mains are again flushed to remove residual chlorine. The flushing waters are dechlorinated by the same techniques used for dechlorinating main flushing waters for maintenance activities.

Chlorinated trench water is dechlorinated by various methods, depending on the circumstances.

The City of Portland, on one occasion, used a solution of sodium metabisulfite to dechlorinate some highly chlorinated water in the process of disinfecting a large transmission main. A portable dechlorination unit similar to that described in Opflow, August 1997 (Eckrich 1997), was used to inject the sodium metabisulfite. The unit was attached to a hydrant steamer port and de-chlorinated about 300 to 500 gpm. Although this de-chlorinating agent was very effective at eliminating chlorine, it was found to be an aggressive oxygen scavenger. Operating engineers also had to wear respirators and other protective gear when handling it. This chemical has not been used since because of its hazardous properties and its deleterious environmental effects.

Portland Water Bureau is in the process of refining its dechlorination practices. Virtually all of the receiving streams that are discharged into are coming under the Endangered Species Act. In the future, the City of Portland will not only have to deal with dechlorination itself, but will also have to address the effects of the dechlorination process.

Detection of dechlorination efficiency. ODEQ guidelines state that the total chlorine residual must be 0.1 mg/L or less before discharge to a receiving stream. Residuals are measured using a Hach pocket colorimeter or Hach color disc test kit (the latter is being phased out). Pocket colorimeters are considered to be accurate for chlorine concentrations up to 0.1 mg/L.

U.S. Utilities Practicing Limited Dechlorination

Tacoma Public Utilities, Tacoma Water (Washington)

Among the participating utilities Tacoma Water Works dechlorinate selected potable water discharges. Dechlorination management practices, target discharges, dechlorination goals and other related information are presented below.

Target waters for dechlorination. Planned releases resulting from disinfection of new mains are dechlorinated. The chlorine concentration in the disinfection water is about 50 mg/L.

Waters not dechlorinated. All unplanned releases from distribution system operations and planned releases, other than new main disinfection water, are not currently dechlorinated. Dechlorination for these waters is currently being evaluated due to the new Endangered Species Act.

Dechlorination agents. Tacoma Water uses sodium metabisulfite powder for dechlorination of new main disinfection waters. Recently, Tacoma Waters has started using sodium ascorbate for dechlorination.

Dechlorination goals. To reduce residual chlorine concentration to 0.0 mg/L.

Dechlorination methods. The main disinfection process followed by Tacoma Water results in a residual chlorine concentration of approximately 50 mg/L. To avoid shocking the sewer with highly chlorinated water, disinfection water is run through a box containing metabisulfite prior to discharge into sanitary sewers. If the water needs to be discharged into a storm sewer, the disinfected water is passed through a treatment trailer where a diluted dechlorinating chemical is added prior to discharge.

Monitoring dechlorination efficiency. Residual chlorine concentration is measured by the Hach colorimetric method.

Dechlorination Practices of Utilities not Participating in This Program

Dechlorination practices of some utilities that are not participating in this program were obtained informally during this reporting period. Although the names of these utilities are not included in this report, their dechlorination practices are provided for the benefit of the reader.

A Utility in Northern California

A certain utility in Northern California follows these dechlorination procedures:

Waters requiring dechlorination. The utility dechlorinates all planned water releases.

Dechlorination agent. Sodium metabisulfite is the primary dechlorination agent used by the District.

Dechlorination methods. For dechlorination of all potable water discharges, except for unusually large flows, the utility uses dry sodium metabisulfite mixed with water at a ratio of 2 ounces of dry chemical to 5 gallons of water to produce a 0.3% by volume solution. The solution is mixed in a 5-gallon plastic container with an adjustable tap on the side just above the base (i.e., a carboy equipped with a spigot). The container is typically placed on the curb above the flow path. The dechlorination solution flow is adjusted using the spigot valve to achieve effective dechlorination, as measured using a total chlorine residual analyzer at a site downstream of the chemical feed point. For larger flushing flows, the utility utilizes a dechlorination trailer rig equipped with a 100-gallon dechlorination tank and dechlorination solution feed pumps.

Monitoring dechlorination efficiency. Chlorine concentrations in treated waters are measured using a Hach DPD analyzer. A redox meter is used to monitor the effectiveness of the dechlorination during large main flushing activities.

pH adjustment. The utility's water ranges in pH from 8.5 to 9.5. No pH adjustment is performed prior to discharge.

A Second California Utility

Another utility in California uses the following dechlorination practices.

Waters requiring dechlorination. The utility requires dechlorination of all planned water releases.

Dechlorination agent. Sodium thiosulfate is the primary dechlorination agent used.

Dechlorination methods. The utility dechlorinates in much the same manner as the previous utility. Sodium thiosulfate crystals are mixed with water in a carboy to form a 1% solution. For flushes, the carboy is set on top of the hydrant and the tap opened to deliver the appropriate dose. Four grab samples per day are taken downstream at random flushing locations to spot-check the effectiveness of the dechlorination method.

Monitoring dechlorination efficiency. Chlorine concentrations in treated waters are measured using a Hach DPD analyzer.

pH Adjustment. The pH of water in distribution system ranges from 6.9 to 8.6. No pH adjustment is performed prior to discharge.

A Third California Utility

The following dechlorination practices are used by another California utility:

Waters requiring dechlorination. All planned water releases are dechlorinated by the utility.

Dechlorination agent. The utility primarily uses sodium thiosulfate for dechlorination.

Dechlorination methods. The utility dechlorinates all potable water discharges using crystalline sodium thiosulfate applied in various ways. For all main flushing, sodium thiosulfate crystals are mixed with water to form a 2-10% solution. The solution is applied in one of two ways. For lower flow discharges, a 2% solution is placed in a 10-gallon container fitted with a tap at the base. The container is placed on the curb above the discharge and the tap opened to discharge the smallest continuous stream possible. For higher flows from hydrants, a Venturi device is used to educt solution from a container into the discharge stream. Typically, a fire hose is attached to the discharge end of the Venturi to provide additional mixing prior to release to the ground surface.

Monitoring Dechlorination Efficiency. In all of the above cases, the chlorine residual is measured downstream of the point of chemical addition to verify that the residual is zero. Water quality staff uses Hach colorimeters to measure chlorine residual. Maintenance staff uses the Hach reagents to produce the color reaction, but the color is assessed visually in comparison to a blank instead of using a colorimeter, since only a positive-negative result is needed.

The utility has spawning coho and steelhead salmon in some of the streams within their service area and is particularly sensitive to receiving water impacts. Both species are classified as threatened under the Endangered Species Act. Consequently, they have conducted extensive monitoring to determine the impacts of adding dechlorinating agents to potable water discharges. In all cases measured, they have found no adverse impacts in terms of pH, dissolved oxygen or sulfide concentrations.

pH adjustment. The utility's water ranges in pH from 7.8 to 8.2, within Basin Plan limits, so pH adjustment is not a concern.

Dechlorination of Cooling Water Releases

Dechlorination of one-pass cooling water releases from two industrial facilities in Tennessee and Kentucky are summarized below.

Dechlorination of Cooling Water Discharge at an Industrial Facility at Tennessee

At this plant, once-through cooling water is dechlorinated prior to discharge into receiving streams. There is one large outfall discharging 4.5 MGD, and 40 small outfalls collectively discharging 3 MGD at this facility. The larger outfall is dechlorinated using a 38% sodium bisulfite solution. A Hach Automatic Chlorine Analyzer monitors the chlorine concentration in the treated water. The initial chlorine concentration in the cooling water is about 0.5 mg/L. The discharge chlorine limitation is 0.3 mg/L. The plant personnel indicated that sodium bisulfite must be handled with caution, since it is an irritant and the solution is viscous and hard to handle. Also, sodium bisulfite crystallizes at temperature below 44° C. Hence, the system must be maintained at a temperature above 44 ° C to avoid plugging of the pipes.

The water chemistry at this site allows for nearly 600% over-dechlorination prior to significant oxygen depletion in the receiving stream. However, a recent release of a slug load of bisulfite, caused by a malfunctioning valve, killed some aquatic species in the stream.

This plant uses sodium sulfite tablets to dechlorinate the smaller streams. Automated tablet dispensers are used. The regulatory chlorine discharge limit for the smaller streams is 0.5 mg/L. A Hach field kit is used to monitor the chlorine concentration.

Dechlorination Practices of an Industrial Facility in Kentucky

This plant also dechlorinates one pass cooling water discharges. There are seven outfalls at this facility, collectively discharging 2.5 - 3.0 MGD water. The chlorine concentrations in these waters are approximately 1 mg/L. The regulatory limit for chlorine is 0.014 mg/L.

Sodium thiosulfate is used for dechlorination. The strength and flow rate of dechlorination solution is adjusted such that four parts of sodium thiosulfate are discharged to neutralize one part of chlorine. The solution is stored in polyethylene carboys and discharged using manually adjusted metering pumps. The tanks are refilled once every two or three days.

The dechlorinated water is discharged into a warm water aquatic stream leading to the Ohio River. Chlorine concentrations are monitored using a field amperometric titration unit mounted in a mobile van. Currently, only chlorine concentrations are measured for regulatory compliance. However, extensive biological monitoring was performed earlier, which indicated no harmful effects upon aquatic life from this plant's discharges.

During summer months, the cooling water release constitutes more than 50% of the receiving water flow. Under such low flow conditions, overdosing of thiosulfate may occasionally result in a white color deposit. This deposit may encourage the growth of sulfur oxidizing bacteria and a subsequent decrease in stream pH.

Canadian Utility Practicing Extensive Dechlorination

Greater Vancouver Regional District (GVRD), British Columbia

GVRD follows extensive dechlorination for all potable water releases. It has developed detailed guidelines for disposing chlorinated water from municipal, construction, agriculture and industrial/commercial uses. GVRD has developed Best Management Practices for reservoir cleaning and water main pigging and flushing operations. GVRD has also developed a Generic Emergency Response for Chlorinated Water Spills and a Chlorine Monitoring and Dechlorination Techniques Handbook. Detailed information can be obtained from the GVRD Internet site 'http://www.gvrd.bc.ca/water/chlorlin/index.html'. A summary of GVRD dechlorination practices is presented in this section.

Discharges requiring dechlorination. GVRD dechlorinates all chlorinated water releases. No unplanned release is exempted from dechlorination by default. An emergency response plan has been developed for containing and dechlorinating accidental releases.

Exempted waters. No water discharges are exempted from dechlorination. However, discharge to sanitary sewers, wherever applicable, is practiced as an option to dechlorination.

Dechlorination agent. Sodium thiosulfate is the primary dechlorination agent used.

Dechlorination goals. The goal of the utility is to reduce the chlorine concentration in discharge waters to below 0.002 mg/L.

Dechlorination of pigging and flushing releases. The preferred method of disposing the water is to discharge the flushing/pigging water to sanitary sewer. The organic and inorganic impurities in the wastewater exert a high level of chlorine demand, resulting in rapid dechlorination. However, the following need to be ascertained prior to releasing chlorinated pigging/flushing waters into sanitary sewers:

- Availability of sanitary sewer connection within a reasonable distance of water mains to be cleaned
- Compliance of flushing water quality with the GVRD sanitary sewer use requirement
- Ability of sewer line to receive the volume of water to be discharged
- Permission from the responsible authority to dispose of the water to the sanitary sewer
- Impact of water quality on treatment plant operation

GVRD emphasizes the need to obtain permission from responsible authorities before discharging into sanitary sewers.

If, due to any of the above reasons the water cannot be discharged into the sanitary sewer, GVRD waters are dechlorinated using a sodium thiosulfate solution. Stock solutions are made up on-site, by mixing anhydrous sodium thiosulfate with water. The following steps are followed in this event:

- Equipment and chemicals kit are prepared and mobilized prior to a cleaning event. The kit consists of dechlorinating agent, a record book, a holding tank for dechlorinating solution, an electric generator and mixer, joint sleeves and fabricated fittings to use with tanks and mixers, a balance, a chemical metering pump with a quick-fix-it kit, water to make dechlorination solution, a flow meter, a diffuser, a field chlorine testing kit, plastic goggles, rubber gloves, a particle mask, MSDSs for thiosulfate, and chlorine testing reagent and a list of emergency phone numbers.

- Initially, the total residual chlorine in the discharge water is measured by Hach Color Comparator model CN-D70 DPD method.
- Stock solution for dechlorinating agent is then prepared
- Flow rates of water to be released and dechlorination dose to be applied are evaluated
- The dosing rate of dechlorinating agent is determined
- Dechlorinating agent is then added, the water is mixed well and the flow is controlled to avoid any adverse impacts
- TRC concentration is then monitored to assess treatment efficiency

Dechlorination of reservoir cleaning waters. Prior to undertaking reservoir cleaning operations, the reservoir is drawn down as low as possible using the existing reservoir pumping system. As much water as possible should be removed from the reservoir, to reduce the volume of water that needs to be removed offsite. Care must be taken not to draw sediments into the pumping system during draw down. The walls and pillars are then cleaned with a high water pressure system. The sediments are than allowed to settle. The quantity of water requiring disposal is estimated.

The first option of discharge is to pump this water to a storm drainage, sanitary sewer or to the land if it can be dispensed in an environmentally acceptable manner. Otherwise, the following on-site treatment options are evaluated for suspended solids/chlorine removal.

If there is sufficient room to construct the settling pond either by using an existing low lying area or by constructing an aboveground pond, the water may be discharged into a temporary settling pond or infiltration ponds.

In addition, if a reasonable percolation rate for the soil is available, the water can be discharged to an infiltration pond.

Accidental water releases. GVRD has developed a comprehensive management program for containing accidental chlorinated water releases. The major issues involved are volume and flow rate of the releases, location of nearby fish environments and likely entry points to the receiving environment, ability to control flow and potential chlorine concentration in the released waters.

One major issue during accidental water releases is the evaluation of possibilities to control flow. Efforts must be made to reduce the pressure on the affected main, and if absolutely required, to halt the flow completely. Closing the water main completely may cause potentially serious contamination, warranting disinfection after repair of broken line.

In addition, released water must be contained by some means. Containment provides extra time to evaluate the seriousness of the situation and to provide neutralization options. Sandbags, earth, tarps, plywood, plastic sheets or a backhoe can be used to contain water flow. The water must be trapped behind a berm before it enters a stream using sand bags or heavy equipment to quickly build a berm. Care must be taken to avoid surcharge of soil resulting in flooded basements. Avoid sand or gravel substrates and use preferentially clay or fine soil for holding.

In addition to control and containment, chlorine neutralization methods must be planned. Sodium sulfite tablets as well as sodium thiosulfate crystals can be used. If water cannot be contained using a berm or dam, one or two dechlorination bags (coarse weave plastic or burlap bags) containing sodium sulfite tablets must be placed in the flowpath. Likewise, if the break occurs close to a stream and trapping the water is not possible, dechlorination bags must be placed quickly on the downstream side, close to the break. If necessary, a small amount of sodium thiosulfate crystals can be added directly to the discharge, before it enters the receiving environment.

Once the water is dechlorinated, it must be safely discharged to the environment. Waters containing residual chlorine should not be discharged to receiving streams or storm sewers. Such waters can, however, be discharged, with permission, into sanitary sewers. Since most field chlorine monitoring kits do not accurately measure chlorine concentrations below 0.1 mg/L, GVRD emphasizes that caution must be exercised to ensure proper discharge of these waters.

Chlorine monitoring. Hach DPD pocket colorimeters are used. These kits do not measure below 0.1 mg/L concentration. However, the presence of a lower concentration of chlorine (up to 0.03 mg/L) can be detected, though not quantified by a slight change in color.

Canadian Utility Practicing Limited Dechlorination

Region of Ottawa-Carleton (Ontario)

The dechlorination practices of this utility are described below.

Target waters for dechlorination. At present, the utility practices dechlorination only for planned releases of chlorinated water that might reach natural waterways (creeks, rivers, etc.). To date, this has only been for reservoirs/pumping stations located in "remote" areas of the distribution system, for which there is no sanitary sewer nearby for disposal. Reservoir discharges after disinfection and discharges after cleaning are dechlorinated. The pH of these waters prior to dechlorination is approximately 8.5 and the chlorine concentration ranges from 0-2.0 mg/L (total chlorine). Reservoir volumes vary from 0.2 - 2.0 million gallons.

Exempted releases. All unplanned releases and planned releases discharging to sanitary sewer lines are exempted from dechlorination.

Dechlorination agents used. Powdered (crystals) or diluted sodium thiosulfate is used for dechlorination.

Dechlorination goals. The ministry of Environment (Ontario) sets the allowable concentrations for discharge to less than 0.002 mg/L chlorine. It is not feasible to test for these concentrations in the field.

Dechlorination methods. Thiosulfate powder is directly applied through reservoir hatches, or with reservoir circulating pumps, if available. Sometimes the powder is pre-diluted before dispersing through reservoir hatches. Generally, it is observed that an excess of sodium thiosulfate (50% extra) is required in practice, compared with theoretical stoichiometric amounts.

Monitoring dechlorination efficiency. Chlorine residual (total) is measured by a field colorimeter or by laboratory analysis using amperometric titration. The titration objective is to read <0.002 mg/L (non-detectable) for total chlorine.

Post treatment. Generally, post treatment is not required when low amounts of chlorine neutralizer are added. Occasionally, pH is adjusted using sulfuric acid prior to discharging water when pH values exceed the guideline of 8.5 standard units.

Summary of Utility Dechlorination Practices

Information obtained from utilities indicates a wide variation in the extent of dechlorination performed, the choice of dechlorination chemicals used and the chemical feed methods employed (Table 4.6). Some regional differences are also observed among the utilities evaluated.

Among the utilities evaluated, in the U.S., the utilities in California, Oregon and Maryland are required to dechlorinate most of their planned potable water releases. In Canada, GVRD dechlorinates all water releases and does not exempt unplanned releases from dechlorination. The utility from Washington State in the U.S. and Ontario in Canada dechlorinate select planned releases such as main disinfection, reservoir cleaning and main flushing waters. Utilities from Texas, Florida, Ohio, Illinois and Connecticut practice little or no dechlorination.

The preferred mode of chlorinated water disposal of all the utilities is discharge to sanitary sewers. The major advantages of this method are i) no chemicals are required, ii) high chlorine demand in sewage can neutralize chlorine content in most of the potable water releases and, iii) the water is not directly released to receiving streams. However, discharge to sanitary sewers must be done with permission and coordination with sanitation agency personnel to ensure safety of workers who may be working in the sewer lines, and to avoid upset to the treatment plant operations.

If discharge to sanitary sewers is not viable, utilities use various chemical agents for dechlorination. Most utilities in California, the GVRD, Region of Ottawa-Carleton and the facility in Kentucky use sodium thiosulfate for dechlorination. Portland Water Bureau uses sodium thiosulfate for selected operations. Sodium thiosulfate is not hazardous, not toxic to aquatic species and does not scavenge oxygen. The feed rate can be better controlled while using this solution. Hence, it is the preferred choice of these utilities.

Calcium thiosulfate has all the benefits of sodium thiosulfate and a slightly more favorable hazard rating. However, it is not used by any of the participating or PAC member utilities. The reason may be due to its late entry into the dechlorination market and slightly higher chemical cost compared to sodium thiosulfate.

EBMUD and WSSC use sodium sulfite tablets for most of their dechlorination needs. GVRD and Portland Water Bureau also use sodium sulfite to dechlorinate select water releases. The industrial facility at Tennessee uses these tablets for dechlorination at smaller outfalls. The advantages of using sodium sulfite tablets are that they are i) easy to store and transport, ii) less susceptible to dust problems and, iii) less likely to splash. However, their dissolution rate may not be uniform throughout application, resulting in minor variations in the degree of dechlorination.

Two utilities use sodium metabisulfite powder for dechlorination. Portland Water Bureau used metabisulfite on one occasion and observed a significant off-gassing problem. Metabisulfite is also reported to scavenge significant concentrations of oxygen.

The industrial facility at Tennessee uses sodium bisulfite for dechlorinating one-pass cooling waters. The personnel indicated problems with crystallization of this agent at room temperatures. Also, an accidental release resulted in a high mortality rate of aquatic species. EBMUD has used sodium bisulfite solution for certain dechlorination activities, however, with no adverse environmental impacts.

Most utilities using dechlorination agents in solution use polyethylene carboys with spigots for feeding the chemicals. The facility at Kentucky and the GVRD use chemical metering pumps for delivery. Those utilities using tablets or powder use burlap or nylon mesh bags to hold the dechlorinating agent. WSSC recommends nylon bags over burlap bags for better results. In addition, WSSC and EBMUD use specially designed devices to regulate contact with the dechlorination chemicals (Chapter 5). Some facilities use automated tablet feeders for dispensing the dechlorinating chemical.

Almost all of the water utilities reported in this section use Hach pocket colorimeters for monitoring chlorine concentrations. The facility at Tennessee uses Hach automatic chlorine analyzers. The industrial facility at Kentucky uses the amperometric titration method for measuring chlorine concentration.

Finally, while working procedures have been developed by several utilities for dechlorination of distribution system waters (1 - 2 mg/L residual chlorine), more work is required to develop procedures for dechlorination of super-chlorinated waters. Super-chlorinated waters (4-50 mg/L residual chlorine) are generated during disinfection of new or repaired main as recommended by AWWA Standard Methods (615-653) for disinfection of distribution mains, storage tanks and reservoirs (Chapter 3, Table 3.1). WSSC currently disposes superchlorinated water by discharging into sanitary sewers at 0.25 MGD. If such an arrangement is not available, the water is discharged to sediment control pond for dechlorination, if available or hauled by contractors for safe disposal. Many utilities dispose super-chlorinated water, either internally or through contractors, by methods as provided for information in AWWA Standard Methods C651, C652 and C653 (Table 4.4).

Table 4.6

Summary of utility dechlorination practices

Utility	Target water for dechlorination	Dechlorination agents	Dechlorination technique	Monitoring methods
EBMUD	All waters except raw water without chlorine residual; groundwaters; and flows < 1000 gallons; unplanned, unmanned releases; and discharge to land.	Sodium sulfite tablets	Discharge into sanitary sewer is the preferred method. For discharges from trenches during main breaks, dechlor mats or strips are used. For hydrant releases, tablets are placed in diffuser chambers.	Pocket colorimeter with DPD reagent.
WSSC	All planned releases except flow and meter test waters.	Sodium sulfite tablets	Prefer discharge into sanitary sewer. Tablet/ suspension, tablet/ diffuser, tablet/ plastic pipe, and tablet/ cage methods are used. Superchlorinated waters are discharged into sanitary sewers or onsite sediment control basins. Else, it is hauled by contractor and dechlorinated.	Pocket colorimeter with DPD reagent.
Portland Water Bureau	Planned releases not discharged into sanitary sewers. Unplanned discharges when there is time to react. If discharge is < 500 gallons or receiving stream flow > 50 ft.^{3}/sec dechlorination may be exempted.	Sodium thiosulfate and sodium sulfite tablets.	Discharge to sanitary sewer is the preferred method. Sodium sulfite tablets in burlap bags are placed in the flow path. For smaller flows, tablet feeders are used. Super chlorinated waters are pumped to tankers, dechlorinated with sodium thiosulfate prior to discharge	Pocket colorimeter with DPD reagent.

(continued)

Table 4.6 (Continued)

Utility	Target water for dechlorination	Dechlorination agents	Dechlorination technique	Monitoring methods
Tacoma Public Utilities	Planned releases from disinfection of new mains.	Sodium metabisulfite/ sodium ascorbate	Main disinfection water (~50 mg/L chlorine) discharged to sanitary sewers is run through a box containing sodium metabisulfite. If discharged to storm sewer, it is passed through a trailer where a diluted treatment chemical is added.	Pocket colorimeter with DPD reagent.
City of Naperville	Main disinfection waters	None	Discharges main disinfection water to sanitary sewers.	None
GVRD	All waters are dechlorinated. No unplanned water is exempted by default.	Sodium thiosulfate, sodium sulfite tablets.	Discharge to sanitary sewers is preferred. Else, sodium thiosulfate solutions are released at rates sufficient to neutralize chlorine. For smaller streams, sodium sulfite is used with tablet feeders.	Pocket colorimeter with DPD reagent.
Region of Ottawa-Carleton	Planned releases that might reach natural waterways. Reservoir discharges after disinfection or cleaning are dechlorinated.	Sodium thiosulfate	Sodium thiosulfate powder is directly applied through reservoir hatches or circulating pumps. Sometimes, the powder is pre-diluted prior to discharge.	Pocket colorimeter with DPD reagent.

(continued)

Table 4.6 (Continued)

Utility	Target water for dechlorination	Dechlorination agents	Dechlorination technique	Monitoring methods
Utility in Northern California (1)	Planned releases	Sodium metabisulfite	For smaller flows, 2 ounces of dry chemical is mixed with 5 gallons of water and is stored in a 5-gallon plastic container with an adjustable tap. The container is placed on the curb above the flow. The solution flow rate is adjusted to ensure total chlorine removal. For large flows, a dechlorination trailer rig with a 100-gallon tank and solution feed pump is used.	Pocket colorimeter with DPD reagent. A redox meter is used to measure the effectiveness during large main flushing activities.
Utility in Northern California (2)	Planned releases	Sodium thiosulfate	Sodium thiosulfate is mixed in a carboy to form a 1-% solution. During main flushing activities, carboys are placed above the hydrant and chemical flow adjusted using the spigot to ensure dechlorination.	Pocket colorimeter with DPD reagent.
Utility in Northern California (3)	Planned releases	Sodium thiosulfate	For low flow, a 2% sodium thiosulfate solution in a container is placed in a curb above the flow.	Pocket colorimeter with DPD reagent.

(continued)

Table 4.6 (Continued)

Utility	Target water for dechlorination	Dechlorination agents	Dechlorination technique	Monitoring methods
Utility in Northern California (3) (continued)			Feed rate is adjusted to ensure dechlorination. For higher flow from hydrants, a venturi device is used to educt solution. A fire hose is attached to the hydrant to allow for mixing. For other flows sodium thiosulfate in a nylon bag is placed on the flow path.	
Industrial facility at Tennessee	Cooling water releases	Sodium bisulfite, sodium sulfite tablets.	A 4.5 MGD flow is dechlorinated using 38% sodium bisulfite solution, with auto feeder and monitoring system. Sodium sulfite tablet feeders are used for 40 smaller flows amounting to 3 MGD.	Automatic chlorine analyzers and pocket colorimeters with DPD reagent.
Industrial facility at Kentucky	Cooling water releases	Sodium thiosulfate	Solution is prepared such that four parts of sodium thiosulfate is delivered to neutralize one part of chlorine. Solution is stored in polyethylene carboys and discharged using manually adjusted metering pumps.	Field amperometric titration unit mounted on a mobile van.

CHAPTER 5

NEW TECHNOLOGIES

INTRODUCTION

In this section, recent developments in the field of dechlorination reported by utilities and industry are presented. Three chemicals (ascorbic acid, sodium ascorbate and hydrogen peroxide) have been recently identified that effectively neutralize chlorine. In addition, two media (catalytic carbon and copper-zinc redox process) are reported to remove residual chlorine from waters. New techniques for delivering dechlorination chemicals into chlorinated waters have been developed by various agencies. Furthermore, a water quality test strip for measuring chlorine using 3,3', 5,5'-tetramethylbenzidine (TMB) has been developed by a company. Applications of these chemicals, techniques, feed methods chlorine-monitoring method for dechlorination of potable waters are discussed in this section.

DECHLORINATION CHEMICALS

Use of ascorbic acid, sodium ascorbate and hydrogen peroxide for disposal of chlorinated water are described in this subsection.

Ascorbic Acid (Vitamin C)

The December 1998 issue of Opflow contains an article on the use of vitamin C (ascorbic acid) for neutralizing chlorine from potable waters (Peterka, 1998). Vitamin C is reported to react with chlorine to produce chloride and dehydroascorbate. With chloraminated water, it is reported to produce insignificant concentrations of ammonium. The reactions with chlorine and chloramine are shown below:

$$C_5H_5O_5CH_2OH + HOCl \rightarrow C_5H_3O_5CH_2OH + HCl + H_2O$$

Ascorbic acid   Hypochlorous acid   Dehydroascorbic acid   Hydrochloric acid

$$2C_5H_5O_5CH_2OH + 2NH_2Cl \rightarrow 2NH_4^+ + Cl_2 + 2C_5H_3O_5CH_2OH$$

Ascorbic acid   Chloramine   Ammonia   Chlorine   Dehydroascorbic acid

Since ascorbic acid is weakly acidic, the pH of water may decrease slightly. As per this article (Peterka 1998), field application of vitamin C reduces far more chlorine than an equal weight of sulfur-based compounds. On a weight-by-weight basis, one part of ascorbic acid is reported to neutralize ten parts of chlorine.

This report also provides a brief summary on the use of vitamin C for different dechlorination applications. To dechlorinate water from fire hydrants, the use of Venturi injectors or other devices such as the one described in the Opflow August 98 issue are recommended.

The report recommends that care be taken to prevent a drop in pH level below 6.5, particularly during the spawning season or when fingerlings are present. Monitoring of the dissolved oxygen concentration and other water quality parameters are highly recommended.

Ascorbic acid is reasonably stable in a dry state with a shelf life of about three years (Peterka, 1998). However, it rapidly oxidizes in solution. Stability of the solution is affected by concentration, exposure to light and air. A 5% weight/volume solution may remain at approximately 95% potency level after 12 days, if kept in the dark, whereas a 1% solution may remain at approximately 80% potency after 10 days. A 0.02% solution will degrade to 0% within three days.

Use of vitamin C is reported to have other potential benefits as it is an essential vitamin for healthy fish (Peterka 1998). Also, it can easily strip manganese oxide stains from reservoir surfaces and promote better disinfection (once the vitamin C is exhausted).

A utility in the Pacific Northwest conducted preliminary dechlorination studies using ascorbic acid. Results indicated that ascorbic acid effectively removed chlorine from hydrant flows. However, the utility reported that three parts of ascorbic acid was required to neutralize one part of chlorine. Furthermore, in some cases, the pH of the discharge water decreased by about 1.0 unit. In some cases, pH levels in storm drains receiving dechlorinated waters decreased by 0.5 to 1.0 units.

Another utility in Oregon conducted some laboratory evaluations of dechlorination using ascorbic acid. The results indicated that the amount of chemical required for dechlorination varied with the initial concentration of residual chlorine in the water. For initial chlorine concentrations of 0.4 mg/L to 0.67 mg/L, the ascorbic acid required to neutralize 1 part of chlorine varied from 2.35 to 2.9 parts. In addition, the data indicated that, in the unbuffered water, the pH decreased from about 9.5 to as low as 4.6 units.

Note that the Opflow report (Peterka 1998) indicated that one part of ascorbic acid is required to neutralize ten parts of chlorine. However, studies by the Pacific Northwest and the Oregon utilities indicated that three parts of ascorbic acid is required to neutralize one part of chlorine. Field studies conducted for this report (Chapter 6) also indicated that approximately three parts of ascorbic acid is required to neutralize one part of chlorine.

According to the information obtained from a vendor, the shelf life of ascorbic acid powder is about one year. However, once in solution the ascorbic acid decays within a few days.

Sodium Ascorbate

A utility in the Pacific Northwest is currently evaluating sodium ascorbate for neutralizing chlorine from potable water releases. The pH of sodium ascorbate is approximately 7.0. The expected reaction of sodium ascorbate with chlorine is shown below.

$$C_5H_5O_5CH_2ONa + HOCl \rightarrow C_5H_3O_5CH_2OH + NaCl + H_2O$$

Sodium ascorbate + Hypochlorous acid → Dehydroascorbic acid + Sodium chloride + water

Bench tests were conducted using a 14 g/L solution of this chemical in tap waters containing approximately 1 mg/L chlorine. Residual chlorine concentrations were measured after 1, 5, 15 and 45 minutes of chemical addition. Results indicated that nearly 3.3 parts of sodium ascorbate were required to neutralize one part of chlorine. More than 90% of neutralization occurred within the first minute in the bench studies. The solution pH did not drop by more than 0.1 standard unit from the initial level of 7.3 standard units.

Sodium ascorbate is available in crystalline form in 2.5, 25 or 100 kg boxes. The chemical is very stable with a shelf life of at least one year in a dry state, if kept in a cool, dark place. The cost of a 25 kg box is approximately $598.00 ($23.50 per kg). If four or more boxes are purchased, the cost is approximately $508.00 per box. Cost of a 50 kg box is about $14.00 per kg. Hence, the cost of treating 1 million gallons of water containing 1 mg/L of total residual chlorine using sodium ascorbate is approximately $26.90 (if sodium ascorbate was bought in 25 kg box), which is significantly greater than the cost of using other chemicals identified in Table 4.3.

Since sodium ascorbate is more expensive than ascorbic acid and more sodium ascorbate than ascorbic acid is required to neutralize the same amount of chlorine, treatment using sodium ascorbate is significantly more expensive than treatment using ascorbic acid. However, the utility favors the use of sodium ascorbate, since this chemical does not appreciably alter the pH of the discharge or receiving waters.

Hydrogen Peroxide

Hydrogen peroxide is yet another chemical that can potentially neutralize chlorine in solution. It reacts with free available chlorine in solutions with a pH greater than 7 according to the following reaction:

$$HOCl + H_2O_2 \rightarrow O_2 + H_2O + HCl$$

Hypochlorous acid   Hydrogen peroxide   Oxygen   Hydrochloric acid

On a weight-to-weight basis, approximately 0.48 mg/L of hydrogen peroxide is required to neutralize 1 mg/L of free chlorine. In most cases, oxygen produced by peroxide will remain dissolved in solution. However, while neutralizing super-chlorinated water, in closed environments, the solution may effervesce, and provisions must be made to accommodate the O_2 evolved. The reaction is mildly exothermic, liberating 37 Kcal/mole, as opposed to 199 Kcal/mole when using SO_2.

However, hydrogen peroxide reacts very slowly with combined chlorine and hence is not recommended for dechlorination of waters containing combined chlorine residuals.

A 50% solution of hydrogen peroxide costs about $0.345 per lb from local suppliers. Actual costs may vary with the quantity purchased and place of delivery. The cost is generally higher than that of SO_2 and other sulfite based chemicals. Reaction with hydrogen peroxide produces oxygen, which may be beneficial to receiving waters.

One of the concerns with the use of H_2O_2 is that it is very reactive and rated hazardous when the strength is greater than 52%. It is not rated as Hazardous by the Comprehensive Environment Response, Compensation and Liability Act (CERCLA). It does not require a Risk Management Plan (RMP). However, SARA Title III Section 311/312 classifies hydrogen peroxide as an immediate health hazard and a fire hazard. The minimum threshold quantity for reporting is 10,000 pounds. Hydrogen peroxide at concentrations of 20% or greater is rated as an "oxidizer and corrosive" by the United States Department of Transportation (USDOT) and must be labeled accordingly.

Due to its USDOT and SARA regulatory requirements, hydrogen peroxide may not be the best dechlorination choice for field applications. It may, however, be a viable alternative to SO_2 for industrial, water and wastewater treatment plant applications.

NEW TREATMENT MEDIA

A review of literature indicated that dechlorination can be achieved by passing chlorinated water through columns containing catalytic carbon or a copper-zinc redox media. Some details regarding these dechlorination approaches are discussed below.

Catalytic Carbon

Catalytic carbon is produced and marketed by Calgon Carbon Corporation as an effective technology for removing chloramines from waters. Catalytic carbon is developed by an advanced process where the electron structure of bituminous coal-based granular activated carbon is altered. Alteration of the electronic structure provides the carbon a wide range of chemical reactions not found in conventional carbons. One such reaction is neutralization of chloramines according to the following reactions:

$$NH_2Cl + H_2O + C^* \rightarrow NH_3 + H^+ + Cl^- + CO^* \qquad (1)$$

ammonium chloride   catalytic carbon   ammonia   oxygen with catalytic carbon

$$2NH_2Cl + CO^* \rightarrow N_2 + 2H^+ + 2Cl^- + H_2O + C^* \quad (2)$$

ammonium chloride | oxygen with catalytic carbon | nitrogen gas | | catalytic carbon

As shown, chloramine neutralization is a two-stage reaction. In the first reaction, the carbon surface (C^*) reacts with monochloramine to reduce monochloramine to ammonia and oxidize the carbon surface to form surface oxygen groups (CO^*). In the second reaction, the surface oxygenated carbon groups react with monochloramines to produce nitrogen gas. In the process, the surface carbon group (C^*) is regenerated.

The chloramine removal efficiency of catalytic carbon is reported to be an order of magnitude greater than that of conventional activated carbons used for dechlorination. The catalytic activity of carbon can be measured by the rate at which it decomposes hydrogen peroxide. Peroxide number, which represents the time in minutes required to decompose a fixed amount of hydrogen peroxide, is the unit used to represent catalytic activity. Calgon reports a peroxide number for catalytic carbon as 8 minutes, whereas the peroxide numbers for conventional carbon vary from 40 to 120 minutes.

Various factors such as empty bed contact time (EBCT), influent chloramine concentration, particle size, and temperature influence treatment efficiency using catalytic carbon (Bockman 1997; Farmer and Kovacic 1997). In a study using water containing 2 mg/L influent chloramine concentration, an increase in EBCT from 10 to 30 seconds increased the volume of water treated to below 0.1 mg/L chloramine from 250 bed volumes to 11,000 bed volumes. In a different study, reducing the mesh size from 20 x 50 to 30 x 70 increased the bed volumes treated from 11,000 to 28,000 at a 30 second EBCT and 2 mg/L influent chloramine concentration. Also, an increase in influent chloramine concentration from 2 mg/L to 5 mg/L decreased the amount of bed volumes treated from 11,000 to 4,000 at 30 second EBCT.

Chloramine destruction by this catalytic reaction increases with an increase in temperature. Reports indicate that an increase in water temperature from 14 °C to 22 °C increases the bed volume treated from 1,500 to 4,000, for the same EBCT (30 second) and influent chloramine concentration (5 mg/L).

A catalytic carbon system was field tested at Garland Beverage Company in Texas. The company processes 250,000 gallons of influent water every day for production of soft drinks.

The influent to the carbon unit contained 1 to 4 mg/L of chloramine and the treatment goal was to reduce the chloramine concentration to below 0.1 mg/L. The company installed a 250 cubic feet (10,000 lbs.) catalytic carbon system and operated it at a surface loading of 8 gpm/ft^2 with a contact time of approximately 7.5 minutes.

The performance of the unit was monitored by measuring free and total chlorine, temperature and pH. After a year of operation and a throughput volume of 45 million gallons of water, the unit continued to produce an effluent chloramine concentration of less than 0.1 mg/L.

The advantages of this process are that catalytic carbon does not introduce any new chemical into the water, and it is more efficient than conventional carbon treatment. However, there are certain limitations to using catalytic carbon for potable water dechlorination. First of all, this technology is effective for removing chloramines only. Free chlorine cannot be removed by catalytic carbon reaction. Although reactions 1 & 2 indicate that surface carbon is regenerated during chloramine removal, fouling by organic compounds and inactivation of sites due to oxidation by various compounds can limit the life of catalytic sites. Another disadvantage of catalytic carbon systems relative to potable water discharge involves developing cost effective methods for applying these types of systems in the field.

If more information becomes available on the cost, ease of use and safety effects, it will be incorporated in the report.

Redox Alloy Media for Dechlorination

KDF Fluid Treatment, Inc. (Three Rivers, MI) has patented a Copper/Zinc Redox Alloy Media (RAM) process for dechlorination. This process involves reduction of chlorine to chloride ion with concurrent oxidation of Zn to Zn^{2+}. The energy for redox reaction comes from the potential difference of 1.1 volts between the Zn/Cu couple.

The RAM process consists of a filtration unit containing Zn/Cu powder. The water to be dechlorinated passes through the filtration unit where chlorine is neutralized to zinc chloride. Soluble Zn^{2+} is reported to be redeposited on the medium. Once all the elemental Zn is oxidized, the medium is restored by a washing process. Used RAM media can also be sold as a secondary metal alloy.

Both free and combined chlorine can be neutralized by RAM. The free chlorine neutralization reaction is as follows:

$$Zn + HOCl + H_2O \Leftrightarrow Zn^{2+} + Cl^- + OH^- + H_2O$$
zinc　hypochlorous acid　　　　　　　chloride　　water

Neutralization of chloramine occurs at a slightly slower rate by the following reaction:

$$Zn + NH_2Cl + H^+ \Leftrightarrow Zn^{2+} + Cl^- + NH_3$$
zinc　ammonium chloride　　　　　　　　　ammonia

The Cu/Zn medium is reported to effectively remove 95 to 99% chlorine from potable waters. For flow rates of up to 650 gpm, the manufacturer suggests a maximum loading rate of 30 gpm/ft^2, when the influent chlorine concentration is less than 5 gpm. The manufacturer reported that a pound of RAM media can neutralize up to 53,000 mg of chlorine. The cost of the media is approximately $3.00 per pound. The actual cost may vary with the amount purchased and place of delivery. One study by the manufacturers using super-chlorinated water indicated that the Zn/Cu medium effectively removed chlorine from an initial concentration of 75 mg/L to 0.01 mg/L. Once the media is exhausted, the media can be sold as a secondary metal alloy.

The advantage of using the KDF process is that it does not involve chemical addition. Secondly, unlike the catalytic carbon process, RAM can neutralize free as well as combined chlorine. However, there are certain limitations to this process. First, a fraction of zinc, upon oxidation to Zn^{2+}, may be released to the water. Waters containing 1 - 2 mg/L of chlorine typically produce effluent $Zn2+$ concentrations of approximately 0.5 to 0.8 mg/L. Treatment of super-chlorinated water is estimated to produce an effluent containing up to a 5 to 8 mg/L Zn^{2+}. Zinc at these concentrations may be detrimental aquatic life. Second, hydrocarbons and oxidants in water can coat the redox surface, making it less effective. A 3-cycle backwash is recommended to remove the coatings from the media. Removal of chlorine from brine has also been reported to cause medium caking problems. Also, the capital cost of this system is slightly higher than that of dechlorination chemical systems. For these reasons, this method is probably not suitable for dechlorination of waters discharged to receiving streams.

DECHLORINATION AGENT FEED TECHNIQUES

Dechlor Mat/Strip for Dechlorination Using Tablets

EBMUD has developed this technique to facilitate effective contact between the flow and the tablets during dechlorination. For dechlorination of discharges from trenches during main breaks, the tablets are placed inside synthetic mesh fabric pockets sewn together in a grid or line (called a "dechlor strip" or "dechlor mat" respectively) (Figure 5.1 & 5.2). The mat/strip is supplied by Mike's Products Company (Appendix A). The dechlor mat or strip is laid across the flow path or over the storm drain and either weighted down or nailed to the street to keep it in place. As the discharged water flows over and around the tablets, chemical is released as the water contacts the tablets, reacting with and destroying the chloramines. In this method, effective contact between the flow and the tablets is accomplished by insuring that the tablets are well-distributed across the flow path.

Mats are typically 3' x 4' in size and strips are of 3" x 36" in size. If the discharge flow is large and spread out, mats may be easier to use than multiple strips. Mats can also sized to cover storm drain inlets to dechlorinate waters released into the drains. Strips are smaller, take up less space in vehicles and multiple strips can be used to cover larger flows so their convenience and flexibility make them the appropriate choice unless some of the conditions described above are encountered. Dechlorination procedures for releases from trenches are shown in Table 5.1.

Dechlor Diffuser

EBMUD uses a diffuser dechlorinator manufactured by DAVCO Associates (LPD-250 Line Purge Dechlorinator) for dechlorinating discharges from hydrants or blowoffs. The LPD-250 is a rectangular cross-sectional, conical shaped device that is fitted with a cylindrical feed tube that holds up to 11 sodium sulfite tablets (Figure 5.3). Superdechlorinating kits are also available that enable the use of the LPD-250 for dechlorinating water with Cl_2 concentrations up to 300 mg/L. The superdechlorination kits are based on the use of the liquid calcium thiosulfate solution. Procedures for the use of the LPD-250 with potable water discharges from hydrants are shown in Table 5.2.

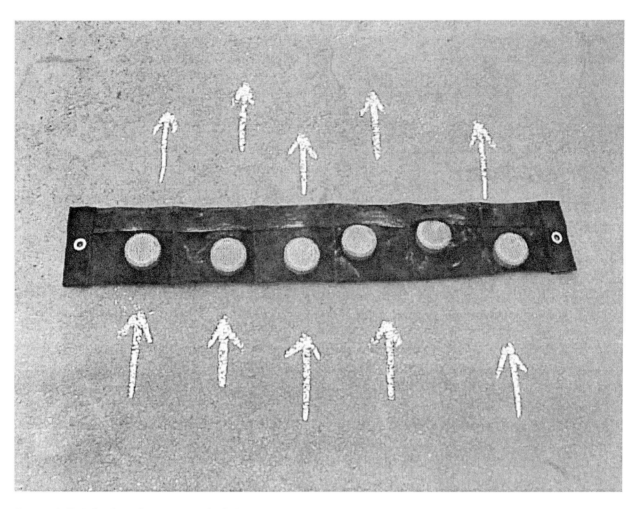

Source: Mike's Products Company, Portland, OR

Figure 5.1 Dechlor strip for dechlorination of trenches during main breaks

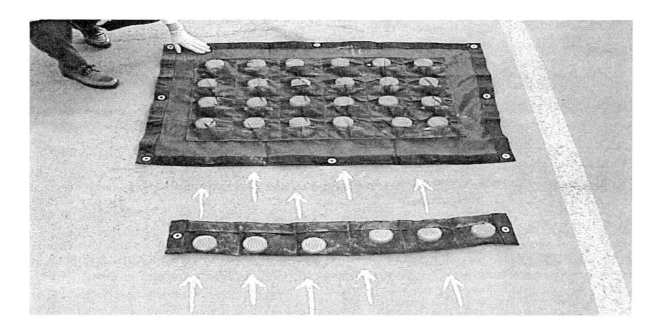

Source: Mike's Products Company, Portland, OR

Figure 5.2 Dechlor mat for dechlorination of trenches during main breaks

Source: Davco Associates, Walnut Creek, CA

Figure 5.3 Diffuser for dechlorination of hydrant or blowoff waters

Table 5.1

Dechlorination procedures for releases from trenches during main breaks

Task	Procedure
1. Fill pockets of dechlor mat or strip with dechlor tablets.	Put one tablet in each pocket of the dechlor mat (3' x 4') or strip (6" x 36"). If the pocket contains a partially used tablet, add another tablet only if there is room.
2. Place dechlor mat or strip mat in flow path	Place the dechlor mat or strip across (perpendicular to) the flow path downstream of sediment control devices (e.g., pea gravel bags). Nail the mat or strip to the street using street nails (through the grommets in either end of the mat) or weigh the mat or strip down to ensure that it stays in place. If the flow path is more than the dechlor mat or dechlor strip, or there is more than one flow path (flow is spreading out in more than one direction), use additional mats to ensure all water from the source is crossing a mat. If the flow is deep (more than 1" above the top of the dechlor mat) and/or the flowrate is very high (>300 GPM), a second mat should be placed downstream of the first mat to ensure adequate dechlorination.
3. Monitor mat or strip	Check the dechlor mat periodically to ensure some tablet remains in each pocket and that all flow is crossing at least one mat.
4. Clean up	When the discharge is complete, move the dechlor mat(s) or strip(s) to the storm drain(s) the discharge was entering, and place it on the upstream side of the grate. Hose the flow path to remove any tablet residual, ensuring that the flow enters the storm drain(s) upon which the dechlor mat(s) or strip(s) is installed. If the flow path separates and some flow travels to a different storm drain, a dechlor mat or strip should be installed at that location as well. Retrieve the dechlor mat or strip and store it in its secondary container on the field vehicle.

Source: East Bay Municipal Utility District, Oakland, CA

Table 5.2

Procedure for dechlor diffuser use

Task		Procedure
1.	Fill diffuser chamber with tablets	Fill the feed tube with at least 10 tablets. If the chamber is partially filled with partially used tablets, add as many new tablets as will fit while still allowing the cap to be screwed back on.
2.	Install diffuser on hydrant, fire hose or blowoff	Screw the LPD-250 to the hydrant, fire hose or blowoff as you normally would. The LPD-250 inlet has a 2.5" female National Pipe Thread. A swivel adaptor will be required to connect to a hydrant or hose. If the diffuser is connected to a hose, be sure that the last few feet of the hose is straight with the diffuser. Also, for flow rates above 600 gpm, a hold down bracket may be required when a hose is used.
3.	Check downstream chlorine residual	Determine if there is any chlorine present downstream, using a pocket colorimeter.
4.	Monitor tablets	Check the supply of tablets in the feed tube after about every 10,000 gallons of flow – replenish as required to fill feed tube.

Source: East Bay Municipal Utility District, Oakland, CA

Tablet Diffuser Dechlorinator

This is a special, patented dechlorination system designed by Washington Suburban Sanitation Commission (WSSC) personnel (Figure 5.4). Permission from WSSC Environmental Engineering and Science section is required for the use of this equipment (Appendix A). The unit consists of a 4.5" circular chamber with sheet metal welded around the sides to create approximately a 6" wide X 3" tall opening. This is a flow-through dechlorination arrangement suitable for discharges from any source that can attach a fire hose connection such as hydrants, pump discharges as well as end-wall type blowoff. The unit holds up to eight Exceltec D-Chlor tablets and can operate up to a flow rate of 120 gpm. The discharge rate can be increased using Siamese adapters or multiple diffusers.

Equipment required for this technique includes: Dechlorination tablets, standard nylon mesh bags cut to 16" X 8", specially designed diffuser, heavy rubber bands, appropriate fire hose/adapters and threaded end wall cap (for end wall discharges only).

The Operation Procedure involves the following:

- Place 8 tablets inside the nylon bags. Push four tablets to each side. Fold the bag over the tablets, twisting in the center. Secure with rubber bands.

- Position the bag firmly inside the diffuser. When discharging from hydrants, attach the diffuser to the 2.5" side of the hydrant using appropriate adapters. Begin discharge, directing the flow to avoid erosion.

- When discharging from blowoff or pump, use a standard fire hose, attach one end of the hose to the threaded end wall and the other end to the diffuser. Position the diffuser so that it does not cause erosion.

Sodium Sulfite Tablet/Bag Suspension Dechlorination

This method is developed and patented by WSSC. This method is suitable for dechlorination of deep manhole type blowoff for a wide range of flow rates. Dechlorination is achieved by suspending nylon bags filled with dechlorination tablets into the manholes (Figure 5.5).

The following equipment is required: Dechlorination tablets, 18 gauge wire, 16"x 32" standard nylon bags, 2" black pipe or similar ones, cut to fit the inside of standard manholes, brackets to secure the pipe in place, nylon rope and sewer bricks for weight.

PROCEDURES

1. Place 8 tablets inside 16"x8" nylon bag. Push 4 to each side as shown. Wear latex gloves and safety goggles.

2. Fold the bag over the tablets, twisting in the center. Secure with rubber bands.

3. Position the bag firmly inside the Porter Diffuser.

4. When discharging from hydrants, attach the diffuser to the 2½ " side of the hydrant using appropriate adapters. Begin the discharge, directing the flow so that it does not cause erosion.

5. When discharging from a blowoff or pump, use a standard fire hose (hose attached to threaded and wall cap pictured).

6. Position diffuser so that it will not cause erosion, using filter cloth or tarp, if necessary.

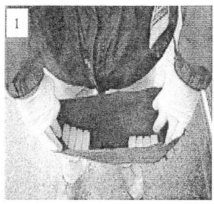
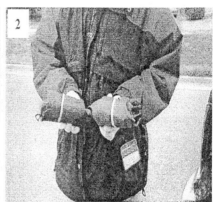
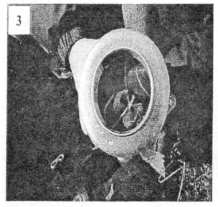

Figure 5.4 WSSC tablet/diffuser dechlorinator

PROCEDURES

1. Fill six standard nylon bags with 20 sodium sulfite tablets in each. Gently place one sewer brick in each bag for weight. Wear latex gloves and safety goggles.

2. Close the bags and secure with wire or rope.

3. Tie 20' to 30' length of strong nylon rope to each bag.

4. Secure a piece of 2" black pipe across the inside of the manhole.

5. Hang 4 of the 6 bags containing the tablets at equal distances along the pipe and at staggered depths.

6. Suspend the remaining 2 bags inside the manhole by either tying them to a nearby structure, such as a tree, or installing a second piece of pipe perpendicular to the first. Make sure that one bag is on either side of the first pipe.

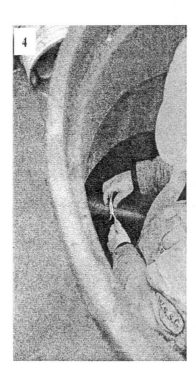
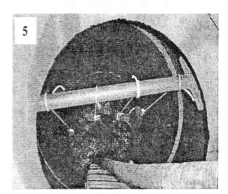
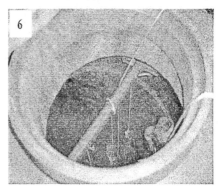

Figure 5.5 WSSC tablet/bag suspension dechlorinator

The Operation procedure for this technique is as below:

- Fill nylon bags with sodium sulfite tablets. Gently place one brick in each bag for weight. Close the bags and secure with wire or rope. Tie 20' to 30' length of strong nylon rope to each bag.

- Secure a piece of 2" black pipe across the inside of the manhole.

- Secure a second 3" black pipe perpendicular to the first pipe.

- Hang the bags containing the tablets at staggered depths.

- A limitation in using this technique is that the direction of flow cannot be controlled.

Tablet/Plastic Pipe Dechlorination

This method is also developed and patented by WSSC. This method is suitable for discharges from end-wall type blowoffs. In this method, nylon bags filled with sodium sulfite tablets are placed inside a 15" corrugated plastic pipe that is secured to an end wall-type blowoff (Figure 5.6).

Equipment required for dechlorination using this technique includes: 16" x 32" standard nylon bags, medium sized stakes, dechlorination tablets, 15" diameter corrugated pipe (specially designed with smaller diameter pipes inside), custom made canvas tube, nylon rope and 18 gauge wire.

The Operation procedure is as below:

- Fill 4 standard nylon bags with about 15 dechlorination tablets each. Secure the bags closed with wire or rope.

- Remove the end wall cap. Tie a 5' to 10' piece of nylon rope to the end wall. Place the small end of the canvas tube over the end wall pipe, pulling the rope through the tube.

- Secure two of the nylon bags along the length of the rope. Position the corrugated pipe around the end wall pipe, inside the canvas tube. Make sure that the two nylon bags are inside the pipe. Secure the pipe to the end wall with rope and stakes.

- Direct the pipe downstream, securing it with ropes and stakes. Make sure that the discharge will not cause erosion.

- Attach the other two bags nylon bags to the outside of the pipe in front of the outlet. Position them to obtain maximum contact with the discharge.

This technique requires a long setup time and large set up area.

Tablet/T -Cage Dechlorination

This technique is also designed and patented by WSSC. Permission from WSSC Environmental Engineering and Science Section is required, for the use of this equipment (Appendix A). This technology is best suited for hydrant flushing and other high velocity discharges. Using this technology, high velocity discharges are dechlorinated using the tablets positioned inside a specially designed metal cage, approximately 3.5' long, 1' wide, 1' tall, with wire mesh at ends (Figure 5.7). The discharge flows through a T which is positioned inside the cage so that the flow is directed to the tablets inside nylon mesh bags secured to the inside ends of the cage.

Required equipment are tablets, a 4" T connection, 18 gauge wire and/or nylon rope, 16" x 32" standard nylon bags, 4" hose and adapters.

The operational procedure is as below:

- Place about 15 tablets in two nylon bags. Secure the bags closed with the rope.
- Secure the closed bags to the cage end walls with wire or rope. Make sure that the bottom seams of the bags are just touching the ground.

PROCEDURES

1. Fill 4 standard nylon bags with 15 sodium sulfite tablets each. Secure the bags closed with wire or rope. Wear latex gloves and safety goggles.

2. Remove the end wall cap. Tie a 5' to 10' piece of rope to the end wall. Place the small end of the canvas tube over the end wall pipe, pulling the rope through the tube.

3. Secure 2 of the nylon bags along the length of the rope.

4. Position the corrugated pipe around the end wall pipe, inside the canvas tube. Make sure that the two nylon bags are inside the pipe. Secure the pipe to the end wall with rope and stakes.

5. Direct the pipe downstream, securing it with ropes and stakes. Make sure that the discharge will not cause erosion.

6. Attach the other two nylon bags to the outside of the pipe in front of the outlet. Position them to achieve maximum contact with the discharge.

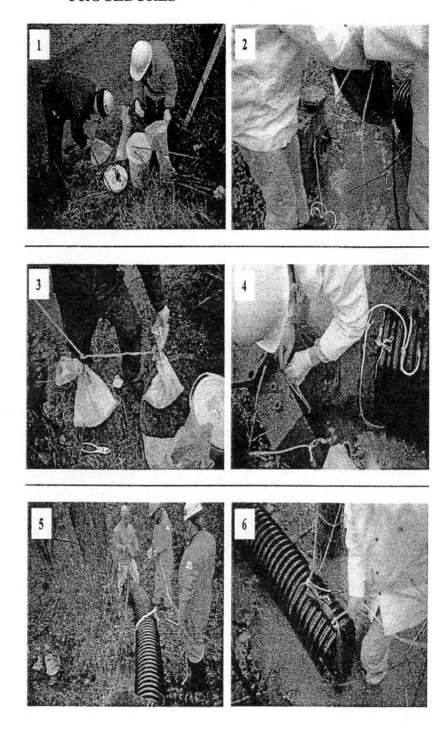

Figure 5.6   WSSC plastic pipe dechlorinator

PROCEDURES

1. For hydrant flushing, place 15 tablets inside 2 standard nylon bags. Secure the bags closed with rope or wire. Wear latex gloves and safety goggles.

2. Secure the bags to the cage end walls with rope or wire. Make sure that the bottom seam of the bags are just touching the ground.

3. For hydrant flushing, attach the 4" hose to the hydrant.

4. Attach the 4" T to the 4" hose.

5. Position the T inside the cage so that the outlets are facing the bags.

6. Begin the discharge. Reposition the T if necessary to achieve maximum contact with the tablets.

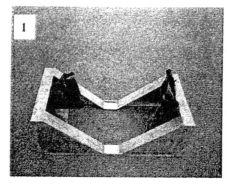

Figure 5.7 WSSC Tablet/T-cage dechlorinator

- Attach the 4" hose to the hydrant. Attach the 4" T to the hose.

- Position the T inside the cage so that the outlets are facing the bags.

- Begin the discharge.

Table 5.3 summarizes the feasibility of each method for waters released from different sources.

Table 5.3

Feasibility of WSSC feed techniques for waters released from different sources

Discharge configuration	Tablet diffuser	Bag suspension	Plastic pipe	Hose monster/ cage*
Hydrant	✓			✓
Deep manhole †	✓	✓		✓
Shallow manhole †	✓			✓
End wall †	✓		✓	✓
Pump †	✓			✓

Source: Washington Suburban Sanitation Commission

* Used only with high velocity discharges and should not be left unattended for more than an hour.

† Diffuser and Hose monster techniques used for these discharges if the blowoff is fitted with a threaded connection or if a connection can be made by the field crew.

Portable Dechlorinator

The August 1997 issue of Opflow (Eckrich, 1997) describes a portable dechlorinator, designed by the City of Blue Springs, Blue Springs, Missouri, for dechlorinating water released from fire hydrants (Figure 5.8). Dechlorination is accomplished by gravity-feeding a sodium thiosulfate solution into the dechlorinator, which disperses the solution into the flow from the hydrant. A list of materials and costs for this technique are presented in Table 5.4.

Table 5.4

Equipment required for construction of portable dechlorinator

Quantity	Material	Cost ($)
1	5-gallon bucket	4.00
1	1/2 in. OD X 2 in. long pipe	2.00
1	1/2 in. ID X 3-ft long rubber hose	3.00
1	1/2 in. male to 3/8 in female adapter	2.00
1	1/2 in. ID X 8-in. long pipe	5.00
1	3/8 in. female to 1/2 in. male adapter	2.00
1	3/8 in. ball valve	4.00
1	4-in. ID X 10 in. long pipe	10.00
1	4-in fire hydrant cap	20.00

Source: Eckrich, J.A. 1997. Portable Dechlorinator Takes Top 1997 Honor. Reprint from Opflow, V.23, No.8, August, by permission. Copyright ©, American Water Works Association.

Source: Eckrich, J.A. 1997. Portable Dechlorinator Takes Top 1997 Honor. Reprint from <u>Opflow</u>, V.23, No.8, August, by permission. Copyright ©, American Water Works Association.

Figure 5.8   Dechlorination chemical feed unit by City of Blue Springs, Missouri

Construction and usage.

1. Weld the 1/2 in. OD X 2-inch long pipe to the bottom of the 5-gallon bucket.

2. Cut four 1/4-inch diameter holes in one side of the 1/2 in. ID X 8-in. long pipe and cap one side.

3. Weld the 8-in long pipe to the 4-in pipe so the holes point towards one opening of the 4-inch pipe.

4. Weld the fire hydrant cap onto the end of the 4-inch pipe opposite the end that the holes of the smaller pipe open toward.

5. Attach the ball valve to the 1/2 in. ID X 8-in. long pipe, which is now welded to the 4-inch pipe.

6. Attach the ball valve to the rubber hose.

7. Attach the rubber hose to the 1/2 in. OD X 2-inch long pipe that is welded to the bucket.

8. Close the ball valve and mix the sodium thiosulfate solution in the bucket. The amount of thiosulfate will depend on the amount of chlorine in the system and the rate at which the solution will be introduced.

9. Attach the hydrant cap to the 4-inch opening on the fire hydrant.

10. Make sure the bucket is at an elevation above the hydrant. Then flush the hydrant in the standard manner that hydrants are flushed, adjusting the ball valve to achieve the rate appropriate for introduction of the sodium thiosulfate solution.

Chlorination/Dechlorination Unit by Water and Wastewater Technologies, Inc., WA

Recently, Water and Wastewater Technologies, Inc. at Bellingham, WA has developed and patented a Chlorination/Dechlorination (C/DC Model 3M-FHA) chemical feeder unit. The C/DC is a well-constructed commercial grade unit that attaches directly to 2 1/2" fire hydrant ports and blowoff assemblies (Figure 5.9). This unit is self-contained and requires no external power or injection pumps. This is a portable unit, smaller in size (20" X 11") and weighs 11 lbs and hence, is convenient for field applications. This unit is reported to dechlorinate waters at flow rates ranging from 20 to 1000 gpm. The dechlorination chemical feed rate can be as high as 38 to 40 gph. The cost of the equipment is approximately $1000.

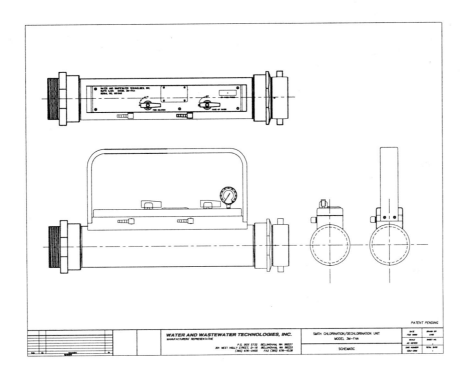

Source: Water and Wastewater Technologies, Inc., Bellingham, WA

Figure 5.9   Dechlorination chemical feed unit by Water and Wastewater Technologies, Inc., Bellingham, Washington

The specification and standards, controls as indicated by the manufacturers are as below:

Specifications of the Equipment

- Overall Length: 20"
- Overall Height: 10"
- Approx. Weight: 11 lbs.
- Flowrate: 20 – 1,000 gallons per minute
- Power: none

Components

- 6061-T6 Aluminum body and machined components
- Venturi block assembly and orifice plate(s)

- 0 – 100 psi pressure gauge
- 2 1/2" female swivel adapter fire hose fitting
- 2 1/2" male hex nipple fire hose fitting
- 8' x 7/16" feed solution suction tubing
- 8' x 7/16" make-up solution discharge tubing
- 38 – 40 gph maximum feed solution suction rate

Controls and Maintenance

- Feed solution control valve
- Make-up solution control valve
- Reference manual, flow and applications data table(s)
- Removable access plates and components for easy maintenance

Operational Description.

Dechlorination of a wide range of flow rate (20 - 1000 gpm) is accomplished with the help of orifice plates of varying sizes provided with the equipment. An orifice for a particular application is chosen based on the estimated flow rate of the chlorinated water. The orifice plate is inserted into the barrel of the dechlorination unit prior to attaching the dechlorination unit to a hydrant port or diffuser assembly. If required, a hose can be attached to the downstream end of the unit to direct the dechlorinated water to a desired discharge point.

The feed solution suction line is inserted into the dechlorination chemical container and the feed control valve is adjusted to desired rate. The hydrant valve is then opened, which facilitates the suction of feed chemical into the flow at a rate controlled by the flow rate of the water and feed control valve setting. Based on the residual chlorine concentration of the dechlorinated water, the feed control valve can be precisely adjusted to provide dechlorination chemical just sufficient to neutralize chlorine and avoid over-dosing.

The instrument is provided with a pressure gauge, to estimate the flow rate of chlorinated water, to facilitate orifice plate selection and initial setting.

The dechlorination unit also has an additional valve and discharge line. This helps in uninterrupted chemical feed through a second line when the chemical in the first container is exhausted.

The equipment is manufactured by Water and Wastewater Technologies, Inc. and distributed by Familan Northwest, Inc., and Ferguson Corporation.

Dechlorination Feeder Unit by City of Salem, OR

Source: City of Salem, OR

Figure 5.10 Dechlorinator feeder unit by City of Salem, OR

The City of Salem, OR has developed a simple suction based dechlorination feeder unit for main flushing activities (Figure 5.10). The unit is made of a PVC 32.25" long, 2.5" diameter main shaft with a with a swivel adapter at one end to facilitate connection with a fire hose and a male NST fitting at the other end. A 1" by pass line is connected with the shaft through a 2.5" x 2.5" x 1" Tee connection. The bypass line has two pressure gauges (2.5", 0-160 PSI), one on each side, a feed control valve to adjust the amount of chemical feed and a feed solution line that can be connected to a carboy with a spigot. The centerpiece of the bypass line is a 1' differential pressure injector. The chlorinated water flow through the main shaft is controlled through a 2.5" gate valve. There needs to be a 10 PSI differential between the inlet and outlet of the injector for proper operation. The amount of dechlorination chemical feed from the carboy can be adjusted by controlling the amount of water diverted to the bypass line, and the feed control valves at the bypass line and carboy. Based on the residual chlorine concentration at the downstream side, the chemical feed rate can be adjusted to provide desired level of dechlorination without over-neutralization. The total height of the dechlorination unit is 12". The city has successfully field-tested this unit for dechlorination of main flushing waters.

CHLORINE MONITORING METHOD

Water Quality Strip Using Tetramethylbenzidine Indicator

Industrial Testing Systems, Inc. at Rock Hill, SC has developed chlorine-monitoring strips for measuring free and combined chlorine at various ranges of concentrations. The technique uses 3,3',5,5'-tetramethylbenzidine (TMB) as an indicator. The color of the strip upon immersion in the water sample is compared with a standard color strip provided by the company to determine the chlorine concentration. The advantages of this strip over the current field monitoring test kits, as reported by the company, are:

- A wide range of chlorine concentrations can be measured. Residual chlorine concentration as high as 750 mg/L can be measured without any dilution.
- The strip is sensitive for measuring chlorine concentration as low as 0.02 mg/L.

Separate test kits are available for measuring free and total chlorine. Test kits are available for measuring various ranges of chlorine concentration. Strips are either packaged in bottles or placed in foil packets. A bottle containing 50 strips cost about $14.00 and 25 foil packets containing two strips each cost approximately $24.00.

CHAPTER 6

FIELD DECHLORINATION STUDIES

INTRODUCTION

One of the objectives of the project is to develop Best Management Practices (BMPs) for various dechlorination activities. However, as discussed in Chapters 1, 4 and 5, procedures for dechlorination of potable waters are still evolving. While chemicals such as sodium thiosulfate and sodium sulfite have been used for some time, some utilities are recently evaluating ascorbic acid and sodium ascorbate for chlorine neutralization. Several dechlorination devices are also being developed. Although limited information on the performance of these chemicals is available, so far no attempt has been made to compare the efficiencies of dechlorination chemicals under identical conditions to determine the chemical of choice for various dechlorination applications.

In particular, the following concerns remain regarding the development of Best Management Practices for dechlorination applications:

- All the dechlorination chemicals have not been tested under identical situations to evaluate dechlorination efficiencies and water quality impacts.
- Many chemicals have not been tested for various chlorinated water releases, feed rates, forms, etc.
- Very little information is available on the disposal of superchlorinated water.
- Very limited information is available on dechlorination of groundwaters.
- Complete information on the toxicity of dechlorination chemicals is not available.
- Dechlorination chemical feed methods have not been optimized for various conditions.

Hence, BMPs for disposal of chlorinated waters have not been developed in this report. Instead, some field studies were performed during this project to obtain preliminary data on dechlorination efficiencies of various chemicals. The field tests were conducted at Tacoma Waters, Portland Bureau of Water Works, and East Bay Municipal Utility District. In these studies, dechlorination efficiencies of the chemicals and other water quality impacts were evaluated. In the Tacoma and Portland studies, a 1% solution of the dechlorination chemicals were introduced into the water released from a hydrant. In the EBMUD study, bags or dispensers containing tablets or powders of dechlorination chemicals were placed in the flow path of hydrant water. In all three sites, the water used for the test originated from surface water sources rather than from groundwaters.

While these tests were designed to provide preliminary information about dechlorination, more detailed laboratory and field studies are required to understand dechlorination under various field conditions. The test conditions used at the three sites are briefly summarized in Table 6.1. The scope of the test at each location and the data obtained are summarized below.

DECHLORINATION STUDIES AT TACOMA WATERS

Background

The amount of various dechlorination chemicals required to neutralize one part of chlorine (stoichiometric) is known (Table 4.1). For example, 1.96 parts of sodium sulfite is required to neutralize one part of chlorine in distilled water. However, the rate and extent of dechlorination may vary with factors such as the nature of reaction of each dechlorinating reagent with chlorine, form of the dechlorinating reagent (solution vs. tablet), rate of mixing, temperature, pH, etc. In addition, reaction rates may vary with the concentration of dechlorination agent added. In order to ensure total chlorine removal, utilities often add higher than stoichiometric amounts of dechlorination agent prior to release of chlorinated water. However, dechlorination agents can potentially impact the receiving water quality by affecting DO, pH, etc.

Table 6.1

Brief summary of field test conditions at the three sites

Test location	Disinfectant used	Chemicals tested	Form of chemicals used	Chemical feed conditions	Initial chlorine conc. (mg/L)	pH	Source of water	Flow rate (gpm)
Tacoma Waters, Tacoma, WA	Free chlorine	Sodium metabisulfite Sodium sulfite Sodium thiosulfate Calcium thiosulfate Ascorbic acid Sodium ascorbate	1% solution	Stoichiometric concentrations and twice the stoichiometric concentrations needed for dechlorination	1.2	8.9	Surface water	300
Bureau of Water Works, Portland, OR	Combined chlorine	Sodium bisulfite Sodium sulfite Sodium thiosulfate Calcium thiosulfate Ascorbic acid Sodium ascorbate	1% solution	Stoichiometric concentrations needed for dechlorination	1.1	8.0	Surface water	300
EBMUD, Oakland, CA	Combined chlorine	Sodium sulfite	Tablets from 2 companies	Varied from 1 to 36 tablets	1.2	8.0	Surface water	100 - 500
		Sodium thiosulfate	Crystals					100
		Ascorbic acid	Powder	1 lb in nylon bags placed on the flow path				100

This field study evaluated the rate of dechlorination of water released from a hydrant at Tacoma Waters facility when the stoichiometric amount of dechlorination agent is added. In addition, the effects of overdosing and concurrent impact on water quality parameters were also evaluated using twice the stoichiometric amount of dechlorination chemicals. Tacoma Waters uses free chlorine for disinfection.

Objectives

The objectives of the field test are:

- To evaluate the following when stoichiometric amount of dechlorination agents are added during a hydrant release event
    - Rate of dechlorination
    - Extent of dechlorination
    - Impact on water quality (pH, DO)
- To evaluate the above when twice the stoichiometric amount of dechlorination agent is added to the released water

Work Plan

Chemical

Sodium metabisulfite, sodium thiosulfate, sodium sulfite, calcium thiosulfate, ascorbic acid, sodium ascorbate in individual solutions of 1% (w/v) were used.

Test Site

The test was conducted at an industrial park in Tacoma. The test was performed in a 450 feet stretch of semi-paved, asphalt road with a hydrant at the upstream end.

Water Quality and Flow Rates

Chlorinated water was released from a hydrant at a rate of 300 gpm. The water had an initial chlorine concentration of 1.2 mg/L. The pH of the water was approximately 8.0. Typical characteristics of Tacoma water are shown in Table 6.2.

Water from the hydrant was released through a fire hose 50 feet long and 6 inches in diameter. The flow rate was controlled using a valve and a flow meter. The other end of the fire hose was connected to a diffuser to facilitate mixing of chemicals with the flow (Figure 6.1). A 1% solution of each chemicals was prepared in a polyethylene bucket and introduced into the flow at the upstream side of the diffuser using an adjustable rate metering pump. The metering pump was pre-calibrated for various flow rates and the flow rate verified prior to each dechlorination test. The feed rate of the 1% (w/v) solution for each of the chemicals needed to neutralize a residual chlorine concentration of 1 mg/L is shown in Table 6.3. Water released from the hydrant traveled approximately 500 feet and discharged into a storm sewer leading to a holding pond.

Table 6.2

Typical water quality characteristics of finished water at Tacoma Water

Compound	Concentration
Alkalinity - As $CaCO_3$	18 mg/L
Ammonia - as N	< 0.01 mg/L
Calcium	4 mg/L
Chloride	< 20 mg/L
Color	5 color units
Conductivity	37 μMHOS/Cm
Free chlorine residual	0.1 to 1.5 mg/L
Hardness - As $CaCO_3$	10 mg/L
pH	7.8 – 8.0
Total dissolved solids	34 mg/L
Total suspended solids	3 mg/L
Turbidity	0.5 NTU

Source: Tacoma Water, Tacoma, WA

Figure 6.1   Field dechlorination test arrangement at Tacoma Water facility

Table 6.3

Chemical feed rate for Tacoma Waters during the field dechlorination studies

Neutralizing chemical	Parts of neutralizing chemical required per part of chlorine (@ pH 8.0)*	Hydrant flow/ chlorine conc (mg/L)	Chemical feed rate for 1% solution (gpm)	
			Stoichiometric dechlorination	100% Over stoichiometric dechlorination
Sodium thiosulfate	1.86	300 gpm (1.0 mg/L)	0.056	0.112
Sodium sulfite	1.96	300 gpm (1.0 mg/L)	0.06	0.12
Sodium bisulfite	1.61	300 gpm (1.0 mg/L)	0.048	0.096
Sodium metabisulfite	1.47	300 gpm (1.0 mg/L)	0.066	0.13
Calcium thiosulfate	1.19	300 gpm (1.0 mg/L)	0.036	0.072
Ascorbic acid	2.48	300 gpm (1.0 mg/L)	0.11	0.22
Sodium ascorbate	2.78	300 gpm (1.0 mg/L)	0.13	0.26

*- Information based on Best Sulfur Product Company brochure and ascorbic acid/sodium ascorbate reactions discussed in Chapter 5.

Sample Collection and Analysis

Samples were collected prior to the point of dechlorination, 2, 100, 200 and 450 feet downstream of the diffuser and analyzed for residual chlorine, DO, pH, redox and turbidity. Upon release from the diffuser, the water traveled for 1 second; 35 seconds; 1 minute, 40 seconds; and 4 minutes, 10 seconds; respectively, to reach the above sampling points. A schematic of field-test arrangements and sampling locations are shown in Figure 6.2.

One of the concerns in introducing dechlorination chemicals is the potential impact of these chemicals upon receiving water quality. It is desirable to limit the amount of dechlorination chemical to that amount which is just sufficient to neutralize the chlorine in the water. Although chemical doses based on stoichiometry were used in this study, the actual field requirements may vary. Hence, the following indirect method was used to determine the amount of dechlorination chemical remaining in the water after dechlorination.

A 5 ml sample of water was collected at the downstream end (450 feet), and its chlorine concentration measured, five minutes after each chemical addition. This water was then added to 5 ml of the hydrant water (which contained no dechlorination chemical). The chlorine concentration in the hydrant water was also measured prior to mixing. In the absence of any dechlorination chemical in the downstream water, the chlorine concentration of the combined water would be the average of the chlorine concentrations in the hydrant and the downstream water. Any decrease in the chlorine concentration of the combined water, below the average concentration, was assumed to be due to presence of dechlorination chemical in the downstream water.

For example, in most cases, the hydrant water contained about 1.2 mg/L of residual chlorine. The residual chlorine concentration of the treated water at 450 feet was typically less than 0.1 mg/L. A decrease in chlorine concentration to less than 0.60 mg/L in the diluted (1:1) sample was considered due to the presence of dechlorination chemical in the treated water.

The residual chlorine concentrations were measured using a HACH pocket colorimeter. According to the manufacturer's specifications, the accuracy of the instrument is within ±0.02 mg/L, but a lower detection limit is not given. Utility experience with this instrument has shown that measurements under 0.1 mg/L are unreliable. Therefore, 0.1 mg/L is considered the practical detection limit in these studies.

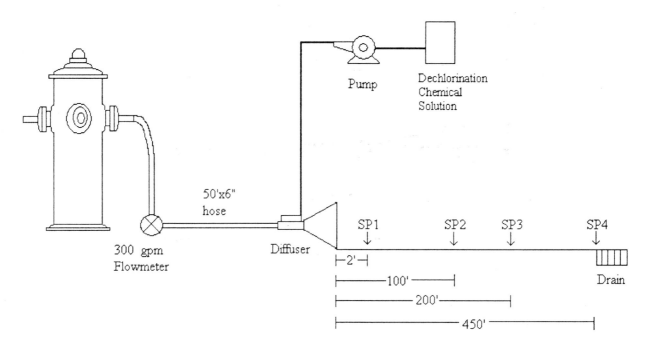

Figure 6.2   Schematic of field test arrangements and sampling points at Tacoma Water Facility

Results

Residual chlorine concentrations. Figures 6.3 & 6.4 show the chlorine concentrations at sampling points after the addition of stoichiometric and twice the stoichiometric concentrations of dechlorination chemicals, respectively. Chlorine concentrations during the flow, when no chemicals were added, are also shown.

The field study indicated that, when no chemicals were added, free chlorine concentration in the water did not decrease significantly. Chlorine concentrations decreased from 1.2 mg/L to approximately 1.0 mg/L after a travel of 450 feet (4 minutes, 10 seconds) in the semi-paved, asphalt road.

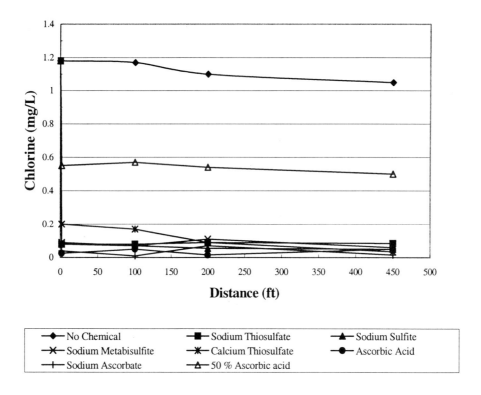

Figure 6.3   Chlorine concentrations at Tacoma City Water when stoichiometric concentrations of dechlorination chemicals were used to neutralize chlorine in potable water from a hydrant

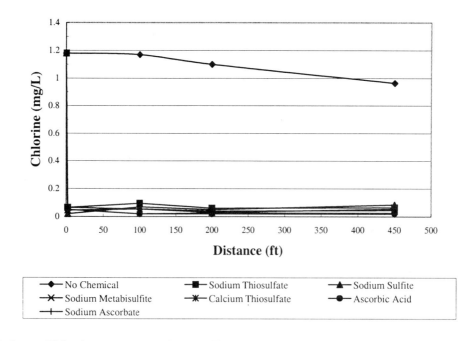

Figure 6.4   Chlorine concentrations at Tacoma Waters when twice the stoichiometric concentrations of dechlorination chemicals were used to neutralize chlorine in potable water from a hydrant

When stoichiometric concentrations of dechlorination chemicals were added, most of the chemicals neutralized chlorine instantaneously. Samples analyzed 2 feet downstream of the diffuser contained less than 0.1 mg/L of chlorine. An exception to this trend was calcium thiosulfate. When calcium thiosulfate was added, chlorine concentrations decreased to 0.2 mg/L within 2 feet and residual chlorine was reduced to less than 0.1 mg/L after a travel of 200 feet.

In one test, due to an oversight in calibration, 50% of ascorbic acid stoichiometrically required to neutralize chlorine was added to the water. Residual chlorine concentrations decreased from 1.2 mg/L to 0.6 mg/L in these samples. This indicated that stoichiometric concentrations of ascorbic acid (2.48 mg of acid to 1 mg of chlorine) were required to neutralize chlorine in the water tested.

When twice the stoichiometric concentration of chemicals was added, residual chlorine in all the tests (including calcium thiosulfate) decreased to below 0.1 mg/L instantaneously (approximately 2 seconds). The trends observed in the two calcium thiosulfate tests indicated that an increase in the amount of dechlorination agent increased rate of chlorine neutralization.

Dissolved Oxygen

The dissolved oxygen (DO) concentration in the hydrant water, prior to the addition of dechlorination chemicals, varied slightly over time (Figures 6.5 & 6.6). During the initial test with no dechlorination chemicals, and in the test using stoichiometric concentration of sodium metabisulfite, the hydrant water contained dissolved oxygen of about 11 mg/L. In subsequent tests using stoichiometric concentrations of various dechlorination chemicals, the DO of the hydrant water, prior to dechlorination chemical addition, was approximately 9.9 mg/L.

Dissolved oxygen concentration of the hydrant water, in the tests using twice the stoichiometric concentrations of dechlorination chemicals was approximately 10 – 10.4 mg/L. The reasons for the fluctuation in dissolved oxygen concentration in the hydrant water, prior to chemical addition, is not known.

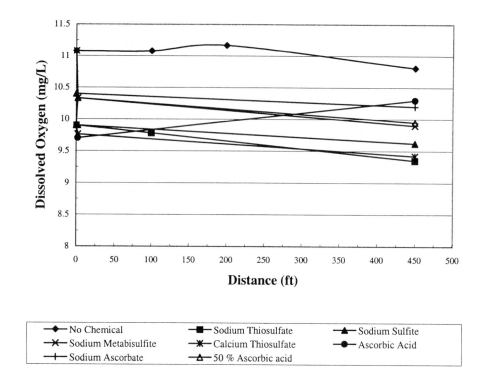

Figure 6.5 Dissolved oxygen concentrations in Tacoma City waters when stoichiometric concentrations of dechlorination chemicals were used to neutralize chlorine in potable water from a hydrant

When no dechlorination chemical was added, the dissolved oxygen concentration of the released water decreased from 11.08 to 10.81 mg/L (0.27 mg/L) after traveling 450 feet in one test, and from 10.4 to 10.3 (0.1 mg/L) in a second test. When stoichiometric amounts of dechlorination chemicals were added, the DO decreased by 1.18, 0.3, 0.55, 0.5 mg/L in the presence of sodium metabisulfite, sodium sulfite, sodium thiosulfate and calcium thiosulfate, respectively, after traveling 450 feet (4 minutes, 10 seconds). The DO increased by 0.3 mg/L in the sodium ascorbate and ascorbic acid tests, after traveling 450 feet. When, twice the stoichiometric amounts of dechlorination chemicals were added, the dissolved oxygen concentration decreased by 1, 0.9, 0.9, 0.7 mg/L in the presence of sodium metabisulfite, sodium sulfite, sodium thiosulfate and calcium thiosulfate, respectively. The DO decreased by 0.2 mg/L in the presence of ascorbic acid and sodium ascorbate.

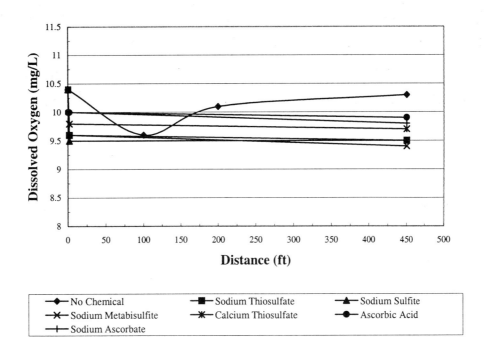

Figure 6.6 Dissolved oxygen concentrations in Tacoma City waters when twice the stoichiometric concentrations of dechlorination chemicals were used to neutralize chlorine in potable water from a hydrant

In summary, results indicated that sodium metabisulfite had a higher impact (1.0 – 1.18 mg/L depletion) on the DO concentrations of the water tested. Sodium sulfite, sodium thiosulfate and calcium thiosulfate decreased the DO concentration by 0.3 to 0.9 mg/L, depending on the amount of dechlorination chemical used. Ascorbic acid and sodium ascorbate had the least impact on the DO of the water tested. The reasons for the increase in DO concentration, in the presence of stoichiometric amounts of ascorbic acid and sodium ascorbate, is not currently known.

pH

The initial pH of the hydrant water, prior to chemical addition, was between 8.8 and 9.0 (Table 6.4). This pH is about 1 unit higher than the target pH of finished water at Tacoma Water. The reason for the elevated pH was later found to be due to an accidental overdose of corrosion control chemicals at a facility just upstream of the test site. Sodium metabisulfite, at either concentration used, decreased the water pH about 0.8 units after a travel of 450 feet.

Ascorbic acid decreased the pH of the water by 0.3 and 0.6 units when stoichiometric and twice the stoichiometric amounts, respectively, were used. In the test using stoichiometric amount of sodium thiosulfate, the pH decreased by 0.4 units. However, no such decrease was noticed when twice this amount of sodium thiosulfate was used. The pH decreased by less than 0.1 units when sodium sulfite, calcium thiosulfate or sodium ascorbate was used at stoichiometric or twice the stoichiometric amounts.

Concentration of Dechlorination Chemicals in the Treated Water

Table 6.5 shows the chlorine concentrations in i) the hydrant water prior to the addition of dechlorination chemicals (hydrant water), ii) water treated with dechlorination chemicals, after a travel of 450 feet (treated water), and iii) sample containing 5 ml each of the above two waters (combined water). If no dechlorination chemicals were present in the treated water, the chlorine concentration in the combined water would be the average of chlorine concentrations in the hydrant and the treated waters (dilution effect). The average of chlorine concentrations for each case is also shown in the table. Results indicated that, in all of the tests, the measured chlorine concentrations of the combined water were significantly lower than the average chlorine concentrations of the hydrant and treated waters. This suggested that, dechlorination chemicals may be present in the treated water, which decreased the chlorine concentration of the combined water below the average concentration. The results indicated the following trends regarding the dechlorination chemicals:

In tests where stoichiometric concentrations of dechlorination chemicals were added, approximately 50 to 80% of dechlorination chemicals remained in the water after 450 feet of travel, although more than 90% of the chlorine was neutralized within 2 feet.

When twice the stoichiometric concentrations of the reagents were added, an amount equivalent to the stoichiometric concentrations of the chemicals remained in the water.

In one experiment where 50% of the stoichiometric amount of ascorbic acid was added, the calculations indicated that one part of chlorine was neutralized by 32% of the theoretically calculated stoichiometric amount of the dechlorination chemicals. However, the chemical neutralized only 50% of the residual chlorine.

Table 6.4

pH of water samples during dechlorination at Tacoma Waters

Chemical	Feed rate	Initial pH	Final pH (450 feet downstream)
No chemical	N/A *	8.8	8.8
Sodium metabisulfite	1 X†	8.8	8.0
Sodium sulfite	1 X†	8.8	8.7
Sodium thiosulfate	1 X†	8.8	8.4
Calcium thiosulfate	1 X†	8.9	8.9
Ascorbic acid	1 X†	8.9	8.6
Sodium ascorbate	1 X†	8.9	8.9
Ascorbic acid	0.5 X‡	9.0	8.8
No chemical	N/A *	8.9	8.8
Sodium metabisulfite	2 X§	8.9	8.0
Sodium sulfite	2 X§	8.9	8.9
Sodium thiosulfate	2 X§	8.9	8.9
Calcium thiosulfate	2 X§	8.9	8.9
Ascorbic acid	2 X§	8.9	8.2
Sodium ascorbate	2 X§	8.9	8.9

*   Not applicable
†   Stoichiometric concentration
‡   50% of stoichiometric concentration
§   Twice the stoichiometric concentration

Although some tests may indicate that less than stoichiometric levels of dechlorination agents can neutralize chlorine, the validity of these data must be verified using detailed laboratory and field testing due to the following reasons:

Due to practical difficulties in the field, the hydrant water with no dechlorination chemical was collected 10 - 15 minutes prior to the collection of dechlorinated test water 450 feet downstream. Exposure of the water to air and sunlight might have affected the results.

Table 6.5
Estimated amounts of dechlorination chemicals remaining in the dechlorinated water at Tacoma

Chemical	Amount introduced	Res. chlorine at the hydrant (mg/L)	Res. chlorine at 450 ft (mg/L)	Avg. of columns 3 + 4	Res. chlorine in combined water (mg/L)*	Estimated amount of dechlorination chemical in the treated water†
Sodium metabisulfite	1 X‡	1.2	BDL§	0.60	0.2	65% of stoichiometric level (0.88 mg/L)
Sodium sulfite	1 X‡	1.2	BDL§	0.60	0.1	83% of stoichiometric level (1.96 mg/L)
Sodium thiosulfate	1 X‡	1.2	BDL§	0.60	0.24	65% of stoichiometric level (1.20 mg/L)
Calcium thiosulfate	1 X‡	1.2	BDL§	0.60	0.24	62% of stoichiometric level (0.73 mg/L)
Ascorbic acid	1 X‡	1.2	BDL§	0.60	0.28	56% of stoichiometric level (1.4 mg/L)
Sodium ascorbate	1 X‡	1.2	BDL§	0.60	0.24	62% of stoichiometric level (1.71 mg/L)
Ascorbic acid	0.5 X**	1.2	0.50	0.85	0.69	32% of stoichiometric level (0.8 mg/L)
Sodium metabisulfite	2 X††	1.2	BDL§	0.60	0.1	86% of stoichiometric level (1.27 mg/L)
Sodium sulfite	2 X††	1.2	BDL§	0.60	BDL§	92% of stoichiometric level (1.8 mg/L)
Sodium thiosulfate	2 X††	1.2	BDL§	0.60	0.16	78% of stoichiometric level (1.45 mg/L)
Calcium thiosulfate	2 X††	1.2	BDL§	0.60	0.11	88% of stoichiometric level (1.0 mg/L)
Ascorbic acid	2 X††	1.2	BDL§	0.60	0.04	93% of stoichiometric level (2.31 mg/L)
Sodium ascorbate	2 X††	1.2	BDL§	0.60	BDL§	91% of stoichiometric level (2.52 mg/L)

* Sample containing 5 ml of hydrant water (prior to dechlorination chemical addition) and 5 ml of treated water after a travel of 450 feet

† Based on the amount of dechlorination chemical required to decrease chlorine concentration in column 4 to the concentration in column five.

‡ Stoichiometric concentration

§ Below detection limit (0.1 mg/L)

** 50% of stoichiometric concentration

†† Twice the stoichiometric concentration

During the field test at Tacoma, a 50 ml graduated cylinder was used to add 5 ml samples of hydrant and dechlorinated water. The volume of water measured by this method may not be very accurate. A small error in measurement may cause a significant difference in the data.

Dechlorination study using 50% of the stoichiometric concentration of ascorbic acid neutralized approximately 50% of the chlorine added, although the dilution experiment indicated that, some ascorbic acid still remained in the treated water (Table 6.5).

In general, results indicated that some dechlorination chemicals may be present in the water after chlorine was neutralized. The potential impact of these chemicals to the water quality, upon complete reaction, was not evaluated in this study. However, as described in the previous subsections (Figures 6.4, 6.5 & Table 6.4), some impact on pH and DO may be observed, if chemical dosage is not controlled during dechlorination.

FIELD DECHLORINATION STUDIES AT BUREAU OF WATER WORKS, PORTLAND

Background

Unlike Tacoma Waters, Portland Bureau of Water uses combined chlorine for disinfection. The field test evaluated dechlorination efficiency and water quality impacts when the stoichiometric amount of dechlorination agent was added. In addition, dechlorination was evaluated when twice the stoichiometric amount of sodium sulfite was added to the water. Experiments with higher doses of sodium sulfite were performed because Portland Water Bureau uses sodium sulfite tablets for most of their dechlorination applications; and rate of dissolution of the tablets may vary during dechlorination resulting in occasional over dosing of the chemicals.

Objectives

The objectives of the field tests are similar to that of the study at Tacoma Water facility. They are:

- To evaluate the following when stoichiometric amount of dechlorination agents are added during a hydrant release event
    - Rate of dechlorination
    - Extent of dechlorination
    - Impact on water quality (pH, DO, ammonia)
- To evaluate the above when twice the stoichiometric amount of sodium sulfite is added to the released water

Work Plan

Chemicals

1% (w/v) solutions of sodium bisulfite, sodium thiosulfate, sodium sulfite, calcium thiosulfate, ascorbic acid and sodium ascorbate were used.

Test Site

The test was performed in a 1000 feet stretch of well-paved asphalt road with a hydrant at the upstream end, in Portland.

Water Quality and Flow Rate

Table 6.6 shows typical water quality characteristics of Portland Bureau waters. Potable water was released from a hydrant at a rate of 300 gpm. The flow rate was regulated using a valve and a flow meter. A 50-foot long, 6-inch diameter hose was connected to the hydrant. The hydrant released the water into a 200-gallon tank in a fire engine. A 1% (w/v) solution of each chemical was delivered to the outlet pipe of the tank using a special feeder developed by the Bureau (Figure 6.7). A 100-foot long, 6-inch diameter hose was connected to the outlet pipe of the tank to allow for the mixing of the dechlorination agent with the hydrant water.

Table 6.6
Typical water characteristics at Portland Water Bureau

Compound	Concentration
Alkalinity - As $CaCO_3$	5.7 - 17 mg/L
Ammonia - as N	ND - 0.34 mg/L
Calcium	1.9 - 2.0 mg/L
Chloride	1.3 - 1.9 mg/L
Color	ND - 15 color units
Conductivity	21 - 43 µMHOS/Cm
Total chlorine residual	0.19 - 2.0 mg/L
Hardness - As $CaCO_3$	5.7 - 6.8 mg/L
Total dissolved solids	20 - 27 mg/L
Total suspended solids	0.5 - 1.5 mg/L
Turbidity	0.09 - 4.6 NTU

Source: Portland Water Bureau, Portland, OR

Sample Collection and Analysis

A schematic of the field-test arrangements and sampling locations are shown in Figure 6.8. Samples were analyzed for residual chlorine concentrations and pH at the hydrant, 2, 100, 500 and 1000 feet downstream of the 100-foot long hose. The travel time for the water to reach the sampling points were 0 minutes, 24 seconds (100 feet); 3 minutes, 2 seconds (500 feet); and seven minutes, 10 seconds (1000 feet). Ammonia concentrations were measured at the hydrant and at the end point (1000 feet). As in the test in Tacoma, the amount of dechlorination chemical remaining in the water after chlorine neutralization was evaluated in this test also.

Results

Residual Chlorine Concentrations

Residual chlorine concentrations are:

Chlorine concentrations at various sampling points are shown in Figure 6.9. When no chemical was added, the chlorine concentration decreased from 1.05 to 0.95 mg/L after

1000 feet. This indicated that only a small amount (0.1 mg/L) of the chloramines dissipated through chlorine demand of paved surfaces. The time elapsed for a travel of 1000 feet was approximately 7 minutes and 10 seconds in this study.

Sodium bisulfite, sodium sulfite, sodium ascorbate and ascorbic acid neutralized all detectable chlorine to below 0.1 mg/L within 2 feet downstream of the mixing hose (approximately 2 seconds).

Sodium thiosulfate neutralized more than 80% chlorine within 2 feet. However, chlorine concentrations decreased below 0.1 mg/L (discharge limit in most states) after about 500 feet (elapsed time 3 minutes, 2 seconds).

Calcium thiosulfate neutralized 60% of the chlorine within 2 feet and neutralized 90% of the chlorine after 1000 feet (elapsed time 7 minutes, 10 seconds).

pH

None of the chemicals used in this study appeared to affect the pH of the Portland Water Bureau water appreciably (Figure 6.10). When no chemicals were added, the pH of the released water decreased from 7.8 to 7.2 after traveling 1000 feet. A marginal increase in the pH was observed (0.3 – 0.4 units) in the tests using sodium thiosulfate, calcium thiosulfate, sodium bisulfite, ascorbic acid and sodium sulfite after traveling 1000 feet. No change in pH was observed in sodium ascorbate test and in the test using twice the stoichiometric amount of sodium sulfite. The alkalinity of Portland water is generally low (5 – 17 mg/L as $CaCO_3$). However, none of the chemicals appeared to impact the pH appreciably, at the concentrations used in this study.

Figure 6.7   Field dechlorination test arrangement at Portland Water Bureau facility

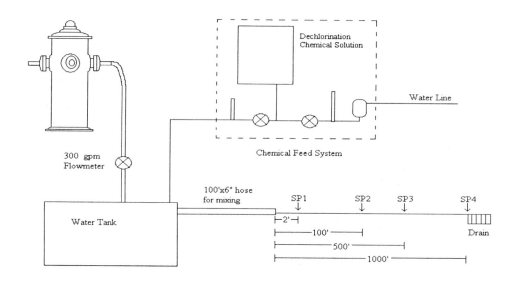

Figure 6.8   Schematic dechlorination test arrangement and sampling points at Portland Water Bureau facility

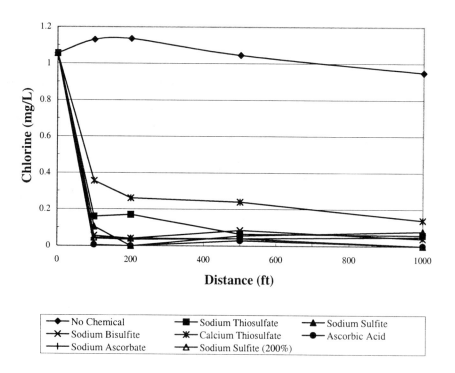

Figure 6.9   Chlorine concentrations of Portland Water Bureau water when stoichiometric concentrations of dechlorination chemicals were used to neutralize chlorine in potable water from a hydrant

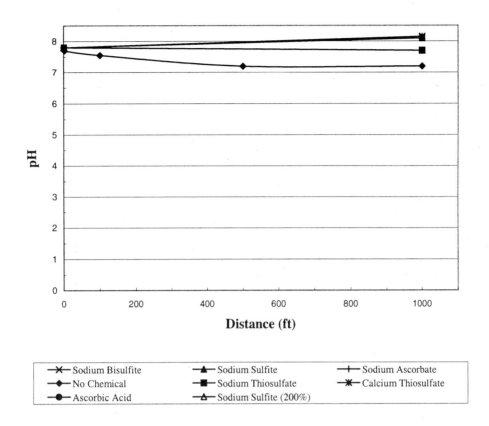

Figure 6.10 pH of Portland Water Bureau water when stoichiometric concentrations of dechlorination chemicals were used to neutralize chlorine in potable water from a hydrant.

Dechlorination Chemicals in the Treated Water

As in the Tacoma study, a 5 ml sample of hydrant water was mixed with 5 ml of dechlorinated water at 1000 feet and analyzed for residual chlorine concentrations. Table 6.7 shows the amount of chlorine anticipated in the diluted water and the amount of chlorine measured. Based on the concentration of chlorine consumed, the concentration of dechlorination chemicals remaining was estimated. It appears that a stoichiometric concentration of each chemical was required to neutralize chloramines at Portland waters.

Table 6.7

Estimated amounts of dechlorination chemicals remaining in the dechlorinated water at Portland

Chemical	Amount Introduced	Res. Chlorine at 1000 ft (mg/L)	Res. Chlorine at the hydrant	Average of columns 3 + 4	Res. chlorine in 1:1 diluted sample*	Estimated amount of Dechlor chemical in the treated water†
Sodium bisulfite	1 X‡	BDL§	1.1	0.55	0.41	30% of stoichiometric level (0.55 mg/L)
Sodium sulfite	1 X‡	BDL§	1.1	0.55	0.36	42% of stoichiometric level (0.9 mg/L)
Sodium thiosulfate	1 X‡	BDL§	1.1	0.55	0.43	25% of stoichiometric level (0.51 mg/L)
Calcium thiosulfate	1 X‡	0.14	1.1	0.62	0.45	31% of stoichiometric level (0.40 mg/L)
Ascorbic acid	1 X‡	BDL§	1.1	0.55	0.64	0% of stoichiometric level (0 mg/L)
Sodium ascorbate	1 X‡	BDL§	1.1	0.55	0.53	9% of stoichiometric level (0.28 mg/L)
Sodium sulfite	2 X**	BDL§	1.1	0.55	0.14	75% of stoichiometric level (1.61 mg/L)

* Sample containing 5 ml of hydrant water (prior to dechlorination chemical addition) and 5 ml of treated water

† Based on the amount of dechlorination chemical required to decrease chlorine concentrations in column 4 to those in column five after a travel of 450 feet

‡ Stoichiometric concentration

§ Below detection limit (0.1 mg/L)

** Twice the stoichiometric concentration

Ammonia Concentrations

The hydrant water contained a free ammonia nitrogen concentration of 0.072 mg/L (NH_3-N) and a total ammonia nitrogen concentration of 0.27 mg/L (Table 6.8). The total ammonia nitrogen concentration after 1000 feet of travel was 0.31 mg/L when no chemical was added. The increase may be due to ammonia present in the impurities on the paved surface. Total ammonia concentrations in sodium bisulfite, sodium sulfite, sodium thiosulfate and calcium thiosulfate neutralized samples after 1000 feet varied between 0.22 to 0.25 mg/L as NH_3-N (0.27 to 0.30 mg/L NH_3). In ascorbic acid and sodium ascorbate neutralized samples, the total ammonia nitrogen concentrations were 0.18 and 0.16 mg/L as NH_3-N (0.22 and 0.19 mg/L NH_3), respectively. The mechanisms of ammonia removal by various chemicals are not currently known. Snoeyink and Suidan (1975) reported that sulfite based dechlorination chemicals do not remove ammonia during dechlorination. However, ascorbate/ascorbic acid appears to remove more ammonia from the water than the other chemicals tested. Further study is required to understand the ammonia trends.

Table 6.8

Ammonia concentrations during dechlorination at Portland

Chemical*	Initial total ammonia concentration (mg/L NH_3-N)	Total ammonia concentration after 1000 feet (mg/L NH_3-N)
No chemical	0.27 (free ammonia 0.072)	0.31 (0.15 free ammonia)
Sodium bisulfite	N/A	0.24
Sodium sulfite	N/A	0.22
Sodium thiosulfate	N/A	0.25
Calcium thiosulfate	N/A	0.24
Ascorbic acid	N/A	0.18
Sodium ascorbate	N/A	0.16
Sodium sulfite (twice the stoichiometric concentration)	N/A	0.23

* - Stoichiometric amount of each chemical was used, unless and otherwise specified.

The ammonia concentrations found in the dechlorinated samples were below USEPA acute toxicity criterion of 0.58 mg/L NH_3 (0.705 mg/L NH_3-N) in the receiving streams under the most conservative conditions (cold water, 30° C, pH 9.0). In addition, a two to four fold dilution of the discharge water prior to discharge into the receiving stream would facilitate compliance with chronic criterion of 0.08 mg/L NH_3 (0.091 mg/L NH_3-N) under the most conservative conditions (cold water, 30° C, pH 9.0).

COMPARISON OF FREE AND COMBINED CHLORINE NEUTRALIZATION

The field test data at Tacoma and Portland indicated potential differences in dechlorination of free and combined chlorine using various chemicals.

Under the test conditions, no significant differences were observed in neutralization of free chlorines and chloramines using sodium bi/metabisulfite, sodium sulfite, sodium ascorbate and ascorbic acid.

However, calcium and sodium thiosulfate appeared to react with combined chlorine at a slower rate than with free chlorine. Sodium thiosulfate neutralized free chlorine to below 0.1 mg/L within 2 feet of the diffuser. However, the combined chlorine concentration decreased to 0.18 mg/L 2 feet away from the hose. Chloramine concentrations decreased to below 0.1 mg/L after 500 feet using sodium thiosulfate.

Calcium thiosulfate decreased free chlorine concentrations to 0.2 mg/L after 2 feet of travel and to 0.1 mg/L after 200 feet of travel at Tacoma and combined chlorine concentrations were 0.36 and 0.21 mg/L, respectively, after 2 and 500 feet at Portland.

In addition to the type of disinfectant used, the water characteristics and experimental arrangement used at the two locations were different. Portland water pH (8.0) was slightly lower than that of Tacoma (8.8). The alkalinity of Portland water was 17 mg/L and the Tacoma water alkalinity is generally around 20 mg/L. A diffuser was used at Tacoma for mixing the chemicals to the water. At Portland, a 100-foot hose was used to provide mixing. These factors may also contribute to the differences in the dechlorination trends observed.

DECHLORINATION FIELD STUDIES AT EBMUD

A field dechlorination study was performed at EBMUD's wastewater treatment facility in June 1998. The purpose of this study was to evaluate dechlorination when chemicals were placed either as tablets or as powder within the path of the chlorinated water. The following dechlorination chemicals were evaluated:

- Exceltech D-Chlor Tablets (91.5% sodium sulfite)
- Norweco Bioneutralizer Tablets (46% sodium sulfite)
- Ascorbic acid (food grade, free white powder)
- Sodium thiosulfate (photo grade 1/8 inch diameter granules)

Experimental Conditions

Six series of field tests were conducted under different conditions. Table 6.9 summarizes the test conditions. Flow was discharged from a hydrant on EBMUD's water distribution system, through a fire hose and onto a fairly level paved and curbed street close to the curb. The water flowed down the street and into a drop inlet 160 feet downstream. The drop inlet led to an onsite storm drain that flows into the headworks of EBMUD's wastewater treatment plant.

Chlorine residual concentrations of the water were measured using a Hach Chlorine Pocket Colorimeter. According to the manufacturer's specifications, the accuracy of the instrument is within ±0.02 mg/L, but a lower detection limit is not given. EBMUD experience with this instrument has shown that measurements under 0.05 mg/L are unreliable. Therefore, 0.1 mg/L is considered the practical detection limit and values of <0.1 mg/L are shown as non-detectable in this report.

A Hydrolab Datasonde equipped with temperature, dissolved oxygen (DO), pH and oxidation-reduction potential (ORP) sensors was used to log real time water quality parameters. The Datasonde was periodically calibrated against standard pH and ORP solutions to verify accuracy.

Table 6.9

EBMUD dechlorination field testing schedule

Test no.	Dechlorination agent	Dosage	Chemical containment method	Flow rate (gpm)
1a	Exceltech D-Chlor	1 tablet	mesh	100
1b	Exceltech D-Chlor	12 tablets	mesh	100
1c	Exceltech D-Chlor	16-28 tablets	mesh	300-500
1d	Exceltech D-Chlor	20 tablets	mesh	50
2a	Exceltech D-Chlor	1 tablet	mesh	100
2b	Exceltech D-Chlor	2 tablets	mesh	100
2c	Exceltech D-Chlor	4 tablets	mesh	100
3	Exceltech D-Chlor	4-20 tablets	feeder	100-500
4	Norweco Bio-Neutralizer	1 tablet	mesh	100
5	Ascorbic acid	1 lb.	fabric	100
6	Sodium thiosulfate	1 lb.	fabric	100

Test Series Descriptions

A brief description of each test series is summarized below.

Test Series I

This series was designed to explore the general equipment arrangement, become familiar with the equipment, measure the effectiveness and evaluate the water quality impacts of dechlorination using Exceltech D-chlor tablets.

Test equipment included:

- Fire hydrant
- Four 50-foot rolls of 3-inch fire hose
- 'Y' fitting to split the flow
- Two rotary paddle type flow meters with gate valves for flow control
- 7x7 per inch polypropylene mesh bags

- D-chlor tablets
- 160 foot long curb-and-gutter section
- Grate covered drop inlet with catch bucket installed by a wire support
- HydroLab Datasonde equipped with ORP, DO, temperature, and pH sensors
- IBM ThinkPad laptop computer
- Hach Chlorine Pocket Colorimeter

The test procedure consisted of placing first 1 D-chlor tablet, then 2, 4, 16, and 28 tablets (in a polypropylene mesh bag) in the water flow 10-feet downstream of the flow control valve and meter set. The flow rates ranged from 100 gpm up to 500+ gpm. Samples were collected 150 feet down stream of the tablets. Water quality data (pH, ORP, temperature and dissolved oxygen) were monitored with the HydroLab Datasonde and collected on the computer. Water samples were collected for laboratory analysis. Chlorine residual measurements were taken by the Hach DPD colorimetric procedure.

Test Series 2

This series was designed to determine if the effectiveness of the Exceltech D-chlor tablets varies with distance from point of chemical addition. Separate tests were run using 1, 2, and 4 tablets in a 100-gpm flow. In this test series, samples were collected at 40-foot intervals along the flow path between the flow control valve and drop inlet. The same equipment used in Test Series 1 was used. Samples were collected at 40-foot intervals in a small plastic beaker. The samples were split; with one portion analyzed for chlorine residual and the other portion analyzed for pH, DO, temperature, and ORP using the HydroLab. For HydroLab readings, the probe was placed in the small plastic beaker as the water was swirled to prevent stagnation. A datapoint was logged after the HydroLab reading had stabilized.

Test Series 3

This series was designed to test the effectiveness of a commercially available tablet feeder (NORWECO) as a substitute for the polypropylene mesh bags used to hold Exceltech D-chlor tablets. The same equipment and test methods used in Test Series 1 were employed except the discharge hose was connected directly to the feeder.

Test Series 4

The test series was designed to measure the effectiveness, evaluate the water quality impacts of dechlorinating using the NORWECO tablets, and to determine if the effectiveness of the NORWECO tablets varies with distance from point of chemical addition. The same equipment and test methods as were used in Test Series 2 were employed.

Test Series 5

Test series 5 was designed to test the performance of ascorbic acid powder under conditions similar to Test Series 1. Rather than using the polypropylene mesh bags for chemical application, two 6-foot long tubes of very fine synthetic fabric were fabricated. One pound of ascorbic acid was poured into the tube. The tube was then crimped with wire ties at 6-inch intervals to ensure the powder remained evenly distributed in the tube upon contact with the discharge flow. One tube per run was placed across the flow, down stream of the control valve and anchored in place.

Test Series 6

The goal was to test the performance of crystalline sodium thiosulfate under the same conditions as Test Series 5.

Test Series Findings

Test Series 1

Residual chlorine concentration. Figures 6.11 - 6.13 and Table 6.10 show results of Test Series 1. As shown in Figure 6.11, one tablet effectively reduced the chlorine residual of the flow it came in contact with, to below 0.1 mg/L for 45 minutes at a flow rate of 100 gpm. Samples for analyses were collected 150 feet downstream of the tablets. The tablet was not fully consumed but became ineffective after approximately one hour. Over the period of exposure, the tablet remained in one piece. It did not break up or become soft. Approximately one-eighth to one-quarter of the tablet remained at the conclusion of this test run.

When 12 tablets were placed across the flow of 100 gpm, the chlorine concentration was below the detection limit after 60 minutes (Figure 6.12). However, when the flow rate was increased to 300/450 gpm, even in the presence of 16 tablets, the residual chlorine concentration increased to values of 0.60 to 0.8 mg/L, well above the detection limit of 0.1 mg/L (which is the allowable discharge limit in many places), within 25 minutes (Table 6.10). This indicates that, under the test conditions used, the flow rate of chlorinated waters can significantly impact the efficiency of dechlorination operations.

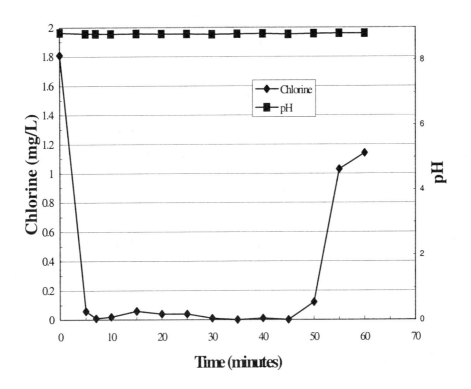

Figure 6.11 Chlorine and pH levels in EBMUD water when 1 sodium sulfite tablet was placed across the flow (100 gpm)

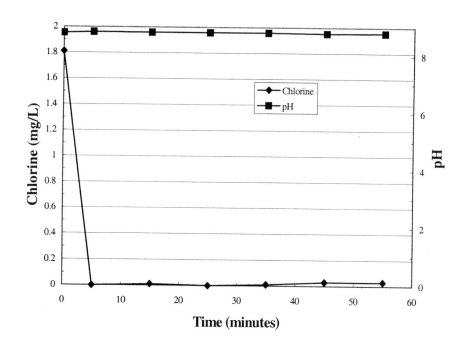

Figure 6.12 Chlorine and pH levels in EBMUD waters when 12 sodium sulfite tablets were placed across the flow (100 gpm)

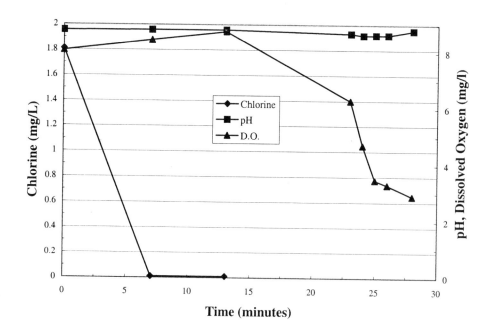

Figure 6.13 Chlorine, DO and pH levels in EBMUD waters when 28 sodium sulfite tablets were placed across the flow (50 gpm)

Table 6.10

Dechlorination using 16 and 20 sodium sulfite tablets in a mesh bag at EBMUD

Time elapsed	No. of tablets	Flow rate (gpm)	pH	DO (mg/L)	Chlorine (mg/L)	Comments
0	16	300	8.75	8.59	1.99	Prior to placing tablets
5	16	300	8.79	9.38	BDL*	
10	16	300	8.79	9.42	BDL*	
19	16	450	8.78	9.11		
23	16	450	8.83	9.79	0.60	
28	16	450	8.84	9.82	0.70	
40	16	450	8.83	9.83	0.80	
45	20	450	8.80	9.48		
51	20	450	8.81	9.83	0.14	
56	20	450	8.82	9.80	BDL*	

* Below detection limit (0.1 mg/L)

Oxidation reduction potential (ORP). The initial oxidation reduction potential (ORP) of the water was about 270 mV. It decreased by approximately 60 units in the presence of one tablet. Once the tablet became ineffective after 45 minutes, the ORP gradually increased to the initial value. With higher doses of sodium sulfite in Test Runs 1b, 1c, and 1d, the ORP shifts were more pronounced. When 12 tablets were placed on the flow path (Test 1b), the ORP decreased from an initial value of 266 mV to 199 mV after 60 minutes. The residual chlorine concentration was below the detection limit (0.1 mg/L) at this time. The test was discontinued at this point. The ORP decreased to as low as 130 mV in the presence of 28 tablets (Test 1c).

pH. No significant impact upon pH was observed in any of the tests. In the test where one tablet was placed across the flow, the initial pH was 8.84. The pH after 60 minutes was 8.80. In test 1b, in the presence of 12 tablets, the initial pH (8.79) did not change appreciably (8.81) after 60 minutes. In test 1c, when 16/20 tablets were placed across the flow (300/450 gpm), the initial pH of 8.75 units did not change by more than 0.1 units after 60 minutes (Table 6.10). In test 1d, when 28 tablets were placed across a flow of 50 gpm, the pH decreased gradually from 8.8 to 8.62 units after 60 minutes (Figure 6.13).

DO. Dissolved oxygen concentrations were not measured during tests 1a and 1b (1 and 12 tablets across a flow of 100 gpm). In test 1c, when 16/20 tablets were placed across a flow of 300/450 gpm, the initial dissolved oxygen concentration was 8.59 mg/L. The DO concentration after 60 minutes was 9.80 mg/L. The DO concentration fluctuated between 9.1 to 9.8 mg/L after the five minutes of the start of the experiment. No specific trend was observed in the DO profile. In test 1d, when 28 tablets were placed across a flow of 50 gpm, the DO concentration decreased significantly from 8.08 mg/L to 2.91 mg/L within 25 minutes. The larger number of tablets and a lower flow rate maintained in this test as compared to the previous three tests, probably caused for the enhanced depletion of DO in the water.

Summary. In summary, results from the test series indicated that, for a flow rate of up to 100 gpm, one Dechlor tablet maintained the residual chlorine concentration below the detection limit for 45 minutes. An increase in number of tablets increased the residual chlorine removal efficiency. However, an increase in flow rate to 300/450 gpm resulted in an increase in residual chlorine concentrations to above detection (and compliance) limits within 25 minutes, even in the presence of 16 tablets. Results also indicated that when the flow rate was decreased (50 gpm) and the number of tablets increased (28) the DO concentration decreased significantly. The pH of the water was not significantly affected under the conditions chosen for this test.

Test Series 2

Residual chlorine concentration. This task consisted of measurements along the flow path, downstream of the point of dechlorinating chemical contact. A flow rate of 100 gpm was maintained, and one or four tablets were placed across the flow. Samples were collected at 40, 80 and 120 feet downstream of the tablets. Results indicated that the residual chlorine concentration in the water decreased with distance. One tablet was sufficient to remove chlorine to below 0.1 mg/L after 120 feet of travel under the test conditions (Figure 6.14).

DO. The DO concentration decreased from 7.0 mg/L to 6.0 mg/L within 80 feet when one tablet was placed across the flow. The DO subsequently increased to 9.60 mg/L after traveling 120 feet. The pH decreased from 8.95 to 8.70 units within 80 feet and decreased to 6.67 units after 120 feet. The reasons for the increase in DO concentration and decrease in pH, after a travel of 80 feet, are not known. However, the trend was less pronounced when four tablets were placed across the flow path (Figure 6.15). The DO concentration decreased from 9.26 mg/L to 9.05 mg/L after a travel of 80 feet. DO then increased to 9.66 mg/L after traveling 120 feet. The pH decreased from 8.95 to 8.9 units after a travel of 80 feet and decreased to 8.75 units after 120 feet. In general, the initial decrease in DO may be due to the reaction of sodium sulfite. Subsequent increase may have resulted from the exhaustion of the released sodium sulfite in the water.

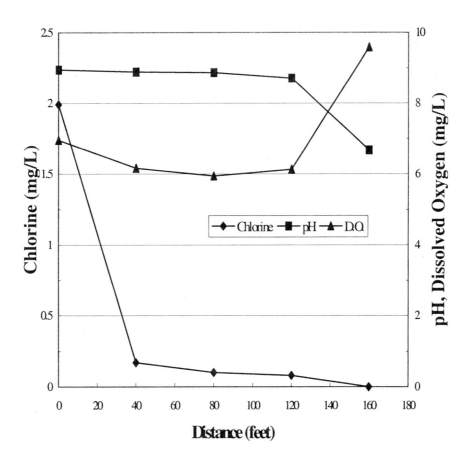

Figure 6.14 Chlorine, DO and pH levels at various points along the flow path using 1 sodium sulfite tablet. The flow rate of the EBMUD water was 100 gpm.

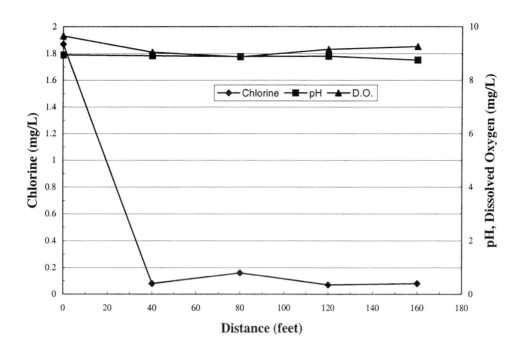

Figure 6.15   Chlorine, DO and pH levels at various points along the flow using 4 sodium sulfite tablets. The flow rate of EBMUB water was 100 gpm.

Test Series 3

Residual chlorine concentration. The test was conducted with a pre-fabricated tablet feeder. A flow rate of 100 or 475 gpm was maintained. Twelve or twenty sodium sulfite tablets were placed across the flow. Samples were collected at 150 feet downstream of the feeder. At a flow rate of 475 gpm, tablet remnants escaped the feeder tubes and were detected downstream of the feeder after the test run. Table 6.11 and Figure 6.16 show the chlorine, DO and pH concentrations during this test. In general, the tablets in the feeder effectively neutralized chlorine under the flow rates (100 - 475 gpm) tested. pH did not change by more than 0.1 unit from the initial value of 8.86.

DO. Some fluctuations were noticed in DO concentrations when a flow rate of 100 gpm was maintained (Table 6.11). Under this flow rate, when 12 tablets were placed in the feeder, the DO concentration decreased from 9.04 mg/L to 8.76 mg/L after 12 minutes. At this point, when the number of tablets in contact with the flow increased to 20 there was a significant decreased in the DO concentration. Within 10 minutes of increasing the number of tablets, the

DO concentration decreased to 8.4. The DO then gradually increased to 8.9 mg/L 25 minutes after increasing the number of tablets.

Similar fluctuations in DO concentrations were not observed when the flow rate was increased to 475 gpm (Figure 6.16). A total of 20 tablets were in contact with the flow in this test. The DO concentration decreased from 8.93 to 8.76 mg/L within one minute and then increased to approximately 9.3 mg/L throughout the duration of the test. A lower reaction time associated with higher flow rate, prior to sample collection may have resulted in the lesser fluctuation in DO at 475 gpm than that observed at 100 gpm.

Table 6.11

Dechlorination of EBMUD water using sodium sulfite tablets in tablet dispensers

Time elapsed	No. of tablets	Flow rate (gpm)	pH	DO (mg/L)	Chlorine (mg/L)	Comments
0	12	100	8.86	9.05	1.98	Prior to placing tablets
12	12	100	8.92	8.76	BDL*	
20	20	100	8.91	8.82		
30	20	100	8.96	8.4	BDL*	
40	20	100	8.94	8.55		
45	20	100	8.92	8.7	BDL*	
53	20	100	8.93	8.91	BDL*	
56	20	100	8.92	8.96	BDL*	
59	20	100	8.91	8.86	BDL*	
64	20	100	8.91	8.9	BDL*	

* Below detection limit (0.1 mg/L)

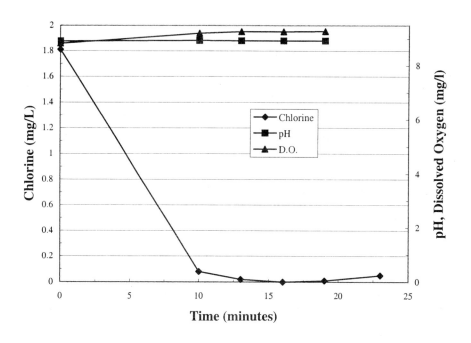

Figure 6.16 Chlorine and pH levels when 20 sodium sulfite tablets were placed in tablet dispensers across 475 gpm flow of EBMUD water.

Test Series 4

Test Series 4 was similar to Test Series 1 and 2 except that NORWECO tablets were used. It quickly became apparent in the initial test that the NORWECO tablets were ineffective at providing full dechlorination under the test conditions. Residual chlorine concentrations decreased from 1.98 mg/L to 0.48 mg/L after a travel distance of 160 feet when one tablet was placed across a flow of 100 gpm (Figure 6.17). Note that, under similar test conditions using Exceltech tablets, the residual chlorine concentration decreased to approximately 0.1 mg/L after a travel of 40 feet (Figure 6.14). A lower content of sodium sulfite (46%) in NORWECO tablet compared to that in Exceltech (91.5%) likely contributed to this difference. In addition, the pH decreased by 1.2 units during this period. Hence, further testing using this tablet was abandoned.

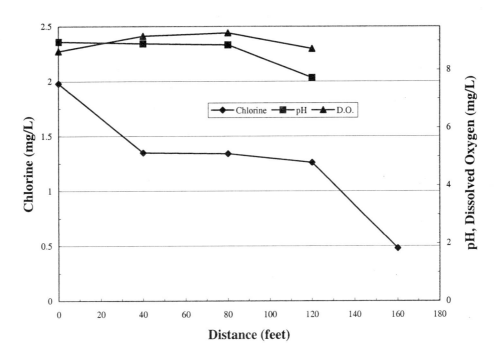

Figure 6.17   Chlorine, DO and pH levels when 1 sodium sulfite tablet (Norweco) was placed across 100 gpm flow of EBMUD water

Test Series 5

This test evaluated dechlorination of EBMUD water using ascorbic acid powder. A flow rate of 100 gpm was maintained. Samples were collected 150 feet downstream of chemical addition. Some problems were encountered due to quick dissolution of ascorbic acid into the flow during the test. Approximately a pound of ascorbic acid powder in nylon bags completely dissolved into the solution within 30 to 60 seconds.

The ascorbic acid depressed the pH from an average for inlet water of 8.9 to 5.07 and the redox potential dropped from an average level of 270 to 116 mV (Table 6.12). The chlorine residual decreased from an average initial concentration of 1.98 mg/L to below 0.1 mg/L, and the dissolved oxygen concentration did not change appreciably. The decrease in pH and ORP may be due to excessive dissolution of ascorbic acid into the water. The lack of decrease in DO concentrations suggested that ascorbic acid may not be a strong reducing agent at the concentrations used in this test.

Test Series 6

Sodium thiosulfate crystals placed in a nylon bag were used for dechlorination in this study. As observed with ascorbic acid, sodium thiosulfate crystals dissolved rapidly when chlorinated water was released at a rate of 100 gpm. The sodium thiosulfate decreased the chlorine residual to below 0.1 mg/L. Sodium thiosulfate had no measurable effect on pH or dissolved oxygen (Table 6.13). Note that in the tests conducted at Tacoma, the dissolved oxygen concentration decreased by 0.55 and 0.90 mg/L when stoichiometric and twice the stoichiometric amounts of sodium thiosulfate were used. This inconsistency in results cannot be readily explained using available data. A 25 mV decrease in ORP was observed during this test.

Table 6.12

Dechlorination using ascorbic acid powder at EBMUD

Parameter	Before chemical addition	After chemical addition
Chlorine (mg/L)	1.98	BDL*
pH	8.9	5.07, 5.57
DO (mg/L)	8.9	9.30, 9.46

* Below detection limit (0.1 mg/L)

Table 6.13

Dechlorination using sodium thiosulfate crystals at EBMUD

Parameter	Before chemical addition	After chemical addition
Chlorine (mg/L)	1.98	BDL*
pH	8.9	8.9, 8.91
DO (mg/L)	8.9	9.34, 9.51

* Below detection limit (0.1 mg/L)

Summary of EBMUD Study

The test indicated that dechlorination is possible with all chemicals (tablets/powder) tested. However, sodium sulfite tablets exhibited more effective dose control than ascorbic acid and sodium thiosulfate powder/crystals.

At a flow rate of 100 gpm, one Exceltech tablet maintained residual chlorine concentration to below 0.1 mg/L for approximately 45 minutes.

An increase in the number of tablets to 12 appeared to maintain the residual chlorine concentrations below 0.1 mg/L for a longer period, at the same flow rate (100 gpm).

However, an increase in the number of tablets to 28, at a flow rate of 50 gpm, significantly decreased the DO concentration (~ 5 mg/L) within 25 minutes. This indicated that caution must be exercised in selecting the number of tablets to be used, in order to prevent DO impacts.

When the flow rate was increased to 475 gpm, the residual chlorine concentration in treated water increased to above 0.1 mg/L (0.60 – 0.80 mg/L) within 25 minutes.

Measurement of DO at various distances from the point of tablet addition indicated an initial decrease in DO concentration followed by an increase.

The initial decrease in DO concentration was more pronounced at a lower flow rate (0.6 mg/L, 100 gpm, 21/20 tablets) than at a higher flow rate (0.15 mg/L, 475 gpm, 20 tablets). Lower reaction time associated with higher flow rate prior to reaching the sampling point may have contributed to this effect.

No significant change in pH level (0.2 to 0.5 units) was observed in most of the tests. However, a decrease of two units was observed in one test using one tablet.

Ascorbic acid powder dissolved rapidly and decreased the pH of the water by three standard units.

Sodium thiosulfate crystals also dissolved rapidly, but did not significantly affect the DO or pH in EBMUD waters. However, a decrease in ORP by 60 mV was observed.

Although residual chlorine concentrations below 0.1 mg/L were achieved in most tests, a measurable zero residual was not obtained consistently, under the test conditions, indicating the need for chemical dose rate optimization.

SUMMARY OF TACOMA, PORTLAND AND EBMUD FIELD TEST RESULTS

The results of the field dechlorination studies are summarized in Table 6.14. The studies yielded the following:

All of the chemicals tested in solution, tablet or powder form were able to neutralize free and combined chlorine to below 0.1 mg/L.

Stoichiometric concentrations of dechlorination chemicals in solution removed more than 90% of residual chlorine.

The DO concentration decreased by 1 mg/L when a stoichiometric amount of sodium metabisulfite was added.

A decrease in DO concentrations (~1.0 mg/L) was observed when twice the stoichiometric amounts of sodium metabisulfite, sodium sulfite or sodium thiosulfate were used.

Dechlorination reactions of calcium thiosulfate were slower than those of other chemicals tested.

Reactions of sodium/calcium thiosulfate with chloramine were slower than those with free chlorine.

While ascorbic acid and sodium thiosulfate solutions neutralized chlorine effectively, when used in powder/crystal form, they dissolved rapidly causing water quality concerns by adversely impacting pH and ORP levels.

Sodium sulfite tablets were very effective in dose control and dechlorination. One tablet was sufficient to dechlorinate 2.0 mg/L of chloraminated water to below 0.1 mg/L for 45 minutes when water was released at 100 gpm.

An increase in the number of sodium sulfite tablets to 28 decreased the DO concentration by 5 mg/L.

An increase in flow rate from 100 to 475 gpm decreased the length of time by more than 20 minutes when residual chlorine concentration was below detection limit (0.1 mg/L). However, the impact on DO concentration was less pronounced at the higher flow rate than at lower flow rates.

While residual chlorine concentrations of less than 0.1 mg/L were measured in several tests, the instrument used and field conditions selected did not yield zero chlorine residual in most cases.

Table 6.14

Summary of dechlorination trends observed in the field tests

Chemical	Form	Amount used	Disinfectant	Flow rate (gpm)	Dechlorination efficiency	Other water quality impacts
Sodium metabisulfite	1% solution	1 X*	Free chlorine	300	Chlorine neutralized to < 0.1 mg/L instantaneously	DO decreased by 1 mg/L
		2 X†	Free chlorine	300	Chlorine neutralized to < 0.1 mg/L instantaneously	DO decreased by 1 mg/L
Sodium bisulfite	1% solution	1 X*	Combined chlorine	300	Chlorine neutralized to < 0.1 mg/L instantaneously	No effect on pH; DO not measured
		2 X†	Combined chlorine	300	Chlorine neutralized to < 0.1 mg/L instantaneously	No effect on pH; DO not measured
Sodium thiosulfate	1% solution	1 X*	Free chlorine	300	Chlorine neutralized to < 0.1 mg/L instantaneously	DO decreased by 0.5 mg/L; no significant effect on pH
		2 X†	Free chlorine	300	Chlorine neutralized to < 0.1 mg/L instantaneously	DO decreased by 0.9 mg/L
		1 X*	Combined chlorine	300	80% dechlorination at 2 feet; neutralization below 0.1 mg/L after 500 ft.	No effect on pH; DO not measured
	Granules	1 lb	Combined chlorine	100	The granules dissolved rapidly; chlorine neutralized to < 0.1 mg/L instantaneously	No effect on pH and DO; ORP‡ decreased by 25 mV

(continued)

Table 6.14 (Continued)

Chemical	Form	Amount used	Disinfectant	Flow rate (gpm)	Dechlorination efficiency	Other water quality impacts
Calcium thiosulfate	1% solution	1 X*	Free chlorine	300	80% dechlorination at 2 feet; neutralization below 0.1 mg/L after 200 ft.	DO decreased by 0.5 mg/L; no significant effect on pH
		2 X†	Free chlorine	300	Chlorine neutralized to < 0.1 mg/L instantaneously	DO decreased by 0.7 mg/L
		1 X*	Combined chlorine	300	80% dechlorination at 2 feet; neutralization below 0.2 mg/L after 1000 ft.	No effect on pH; DO not measured
Ascorbic acid	1% solution	1 X*	Free chlorine	300	Chlorine neutralized to < 0.1 mg/L instantaneously	pH decreased by 0.3 units; DO increased by 0.3 mg/L
		2 X†	Free chlorine	300	Chlorine neutralized to < 0.1 mg/L instantaneously	No significant effect on DO; pH decreased by 0.6 units
		1 X*	Combined chlorine	300	Chlorine neutralized to < 0.1 mg/L instantaneously	No effect on pH; DO not measured
	Powder	1 lb	Combined chlorine	100	Rapid dilution of granules; chlorine neutralized to < 0.1 mg/L instantaneously	pH decreased by 3.8 units; ORP‡ decreased by 100 units; no change in DO
Sodium ascorbate	1% solution	1 X*	Free chlorine	300	Chlorine neutralized to < 0.1 mg/L instantaneously	No significant effect on pH; DO increased by 0.3 mg/L
		2 X†	Free chlorine	300	Chlorine neutralized to < 0.1 mg/L instantaneously	No significant effect on pH and DO
		1 X*	Combined chlorine	300	Chlorine neutralized to < 0.1 mg/L instantaneously	No effect on pH; DO not measured

(continued)

Table 6.14 (Continued)

Chemical	Form	Amount used	Disinfectant	Flow rate (gpm)	Dechlorination efficiency	Other water quality impacts
Sodium sulfite	1% solution	1 X*	Free chlorine	300	Chlorine neutralized to < 0.1 mg/L instantaneously	No significant effect on pH; DO decreased by 0.3 units
		2 X†	Free Chlorine	300	Chlorine neutralized to < 0.1 mg/L instantaneously	DO decreased by 0.9 mg/L; no effect on pH
		1 X*	Combined chlorine	300	Chlorine neutralized to < 0.1 mg/L instantaneously	No effect on pH; DO not measured
		2 X†	Combined chlorine	300	Chlorine neutralized to < 0.1 mg/L instantaneously	No effect on pH; DO not measured
	Tablet (Exceltech)	1 tablet	Combined chlorine	100	Chlorine neutralized to < 0.1 mg/L for 45 minutes	No effect on pH; DO decreased by 1 mg/L after 80 feet, then increased to original value
		2 tablets	Combined chlorine	100	Chlorine decreased from 2 to < 0.1 mg/L after 160 ft.	DO dropped by 3 mg/L in one test
		4 tablets	Combined chlorine	100 - 475	Chlorine decreased from 2 to < 0.1 mg/L instantaneously	No effect on pH; DO decreased by 0.25 mg/L and increased to original value
		12 tablets	Combined chlorine	100	Chlorine decreased from 2 to < 0.1 mg/L instantaneously	No effect on pH; DO decreased by 0.3 mg/L
		20 tablets	Combined chlorine	100	Chlorine decreased from 2 to < 0.1 mg/L instantaneously	DO decreased by 0.6 units; no change in pH
		20 tablets	Combined chlorine	475	Residual chlorine remained below 0.1 mg/L only for 25 minutes	DO did not decrease significantly

(continued)

Table 6.14 (Continued)

Chemical	Form	Amount used	Disinfectant	Flow rate (gpm)	Dechlorination efficiency	Other water quality impacts
		28 tablets	Combined chlorine	50	Chlorine decreased from 2 to < 0.1 mg/L instantaneously	ORP[‡] decreased by 100 mV; DO decreased by 5.2 mg/L
	Tablet (Norweco)	1 tablet	Combined chlorine	100	Chlorine decreased from 2 to 0.5 mg/L after 160 ft.	pH decreased by 1.2 units

*   Stoichiometric concentration
†   Twice the stoichiometric concentration
‡   Oxidation reduction potential

CHAPTER 7

RECOMMENDATIONS FOR FUTURE WORK

INTRODUCTION

The current state of dechlorination practices indicates that additional work needs to be performed under varying conditions to better understand the mechanisms and issues involved. In the past few years, several new dechlorination chemicals have been evaluated for effectiveness. In addition, several utilities and industries have developed dechlorination chemical feed equipment. Media such as catalytic carbon have also been developed for chlorine removal. However, the effectiveness of chlorine neutralization by passive, non-chemical methods is not well documented. In order to develop an industry standard and provide a Best Management Practices manual the following issues must be addressed:

- Passive chlorine removal under various potable water discharge conditions
- Dechlorination efficiencies, economy and water quality impacts of each of the chemicals under identical conditions
- Effectiveness of the chemicals and equipment under various release conditions
- Ease, economy, adaptability and effectiveness of various tablet and dechlorination chemical feeders

Passive Chlorine Removal

Several studies have been conducted to evaluate the fate of free and combined chlorine during transmission in distribution systems. However, the chlorine decay during potable water release scenarios is not well documented. The studies at the Tacoma Waters facility indicated that less than 0.2 mg/L of free chlorine was dissipated through chlorine demand of the impurities in a semi-paved surface. Portland Water Bureau field studies indicated that combined chlorine decreased by about 0.2 mg/L after a travel distance of 1000 ft when no chemical was added. Similar results were observed in some EBMUD studies. All of the above works were conducted at a flow rate of 100 to 300 gpm.

A systematic study should be conducted to estimate free and combined chlorine dissipation by various surfaces for the chlorinated water releases documented in Chapter 3 of this report. This will help the utilities to decide the circumstances under which no chemical dechlorination is required when a sanitary sewer is not available for the discharge of chlorinated waters.

Dechlorination Efficiencies of Various Chemicals/Media Under Identical Conditions

A review of current dechlorination practices by the utilities (Chapter 5) indicates that several chemicals are used by the industries for dechlorination of potable waters. In addition, utilities are constantly modifying their practices as more information becomes available about the dechlorination chemicals. Since dechlorination practices are still evolving, the chemicals and media are not tested for different waters and compared against the other dechlorination agents. A comprehensive study would be useful in evaluating the dechlorination efficiencies of various chemicals under identical conditions to select the best approach for a given situation. The evaluation of the chemicals/media should include the following:

- The rate of dechlorination
- Extent of dechlorination
- Amount of chemical required to neutralize unit amount of residual chlorine
- Amount of mixing required
- Water quality impacts such as effects on pH, temperature, dissolved oxygen, ammonia, etc.
- Effects of overdosing
- Costs of the chemical
- Ease of transport and handling, hazard rating, crystallization, etc.

Effectiveness of Each Chemical Under Different Conditions

The characteristics of the water and the nature of each release can impact the dechlorination efficiency of the chemicals and media. For example, ascorbic acid is reported to decrease water pH during dechlorination by some utilities. However, dechlorination of groundwater, which typically contains a higher concentration of alkalinity, may not be significantly affected by ascorbic acid. Furthermore, sodium ascorbate solutions may be very effective for dechlorination planned releases. However, these solutions are not stable for more than a day or two and cannot be stored for use during unplanned/emergency releases. A method that is very effective for high volume release from hydrants may not be best suited for continuous, low volume releases such as leakage, etc. Hence, the effectiveness of each chemical and media should be tested under varying conditions. This may include the following:

- Low, high and moderate flow rates
- Variation in water characteristics such as alkalinity, turbidity, pH, etc.
- Planned, unplanned and continuous releases
- Normal and superchlorinated waters
- Free and combined chlorinated waters
- Hot and cold weather conditions
- Variation in surfaces with regards to passive dechlorination.

Efficiency of Various Chemical Feeders

Several chemical feeders/techniques have been developed by various utilities and industries. The adaptability and economy of the feeders under various conditions must be evaluated. The following criteria are considered important:

- The range of flow rate handled
- The minimum, maximum chemical feed rates
- Ability to control chemical feed rates

- Maintenance required
- Hazard considerations
- Costs
- Ease of use
- Ease of transportation
- Durability.

Improvement in Analytical Procedures

In addition to the above, more research is needed to improve the efficiency of the current field techniques to measure residual chlorine concentrations. The colorimetric method using DPD reagent, which is widely used by the utilities, is reliable for measurement up to 0.1 mg/L of residual chlorine. The discharge limit and receiving water quality criterion for residual chlorine in many U.S. states and Canadian provinces are much lower than this concentration. The questionable practice of measuring chlorine concentrations below 0.1 mg/L, or a visual observation of no color formation, is being used to determine compliance with regulations. In addition, impurities in the released water may interfere with chlorine analyses. Efforts must be made to overcome these difficulties.

Development of Industry Standard/BMP

Upon obtaining consistent and reliable information on the above issues, a user friendly, Best Management Practices should be developed to assist the water utilities with the disposal of chlorinated waters. The BMPs should address the following:

- Type of release: Planned and unplanned releases
- Source of water: Surface and groundwaters
- Anticipated Residual: Normal and super-chlorinated waters
- Type of disinfectant: Free and combined chlorine
- Turbidity: High and low
- Discharge: High and low volume flows

APPENDIX A

UTILITIES/COMPANIES THAT PROVIDED INFORMATION FOR THIS REPORT

Table A.1

Addresses of agencies whose methodologies are presented in this report

Name of the agency	Address
East Bay Municipal Utility District, California	375 Eleventh Street P.O. Box 24055, MS704 Oakland, CA - 94607, U.S.A.
Washington Suburban Sanitary Commission, Maryland	14501 Sweitzer Lane Laurel, Maryland 20707-5902, U.S.A.
City of Portland, Bureau of Water Works, Oregon	1001 SW Fifth Avenue, Suite 450 Portland, OR 97204, U.S.A.
Tacoma Utilities Department, Tacoma Water, Washington	3628 South 35th Street P.O. Box 11007 Tacoma, WA - 98411-0007, U.S.A.
City of Naperville, Department of Public Utilities, Illinois	400 South Eagle Street Naperville, IL 60566-7020, U.S.A
Greater Vancouver Regional District, British Columbia	4330 Kingsway, Burnaby, British Columbia, V5H 4G8, Canada.
Regional Municipality of Ottawa-Carleton, Ontario	Water Division Ottawa-Carleton Center, Cartier Square 111 Lisgar Street, Ottawa, Ontario K2P 2L7, Canada

Table A.2

Addresses of companies that provided information for this report

Company	Address
Arden Industries, Inc.	4720 Fawn Shingle Springs, CA 95682 Phone: 530 677 3671 Fax: 530 677-3672 <ardenin@directcon.net>
Best Sulfur Products A division of Ag Formulators, Inc.	4704 Via Breeza Modesto, CA 95357 Phone: 800 474-5826 Fax: 209 575-4822 <rlhardison@worldnet@att.net>
Calgon Carbon Corporation	400 Calgon Carbon Drive Pittsburgh, PA 15205 Phone: 800 422 7266
DAVCO Associates	2116 N. Main Street, Suite J Walnut Creek, CA 94596 Phone: 925-934-9333 Fax: 925 934 3544
Exceltec International Corporation	1110 Industrial Blvd. Sugar Land, TX 77478 Phone: 800 621 9189 Fax: 281 240-6762
Familian Northwest	13716 N.E. 177th Place Woodinville, WA 98072 Phone: 425 483-8800 Fax: 425 485-4437
Ferguson Corporation	3280 Market Street San Diego, CA 92102 Phone: 619 515-0300 Fax: 619 239-4727
Hach Company	5600 Lindbergh Drive Loveland, CO 80539 Phone: 800 227-4224

(continued)

Table A.2 (Continued)

Company	Address
Hanna Instruments	Marketed by: Industrial Process Measurement, Inc. Hosica Laboratories Newfoundland, NJ 07435 Phone: 973 697-6111 Fax: 973 697- 5234 <www.instrumentation2000.con>
Industrial Testing Systems, Inc.	1875 Langston St. Rock Hill, SC 29730 Phone: 803 329-9712 Fax: 803 329-9743
KDF Fluid Treatments	1500 KDF Drive Three Rivers, MI 49093-9287 Phone: 800 437-2745 Fax: 800 533-3584
Mike's Products Company	13315 NE Whitaker Way Portland, Oregon 97230. Phone: 503 256-5607 Fax: 503 256-5589
Norweco	220 Republic Street Norwalk, OH 44857-1196 Phone: 419 668-4471 Fax: 419 663-5440 <www.norweco.com>
Orion Research Incorporated	500 Cummings Center Beverly, MA 01915-6199 Phone: 978 232-6000 Fax: 978 232-6015
Water and Wastewater Technologies, Inc.	P.O. Box 5732 Bellingham, WA 98227 Phone: 360 Fax: 360 734-2932

REFERENCES

APHA, AWWA, WPCF (American Public Health Association, American Water Works Association, Water Pollution Control Federation). 1989. *Standard Methods for the Examination of Water and Wastewater.* 17th ed. Washington, D.C. APHA.

AWWA (American Water Works Association). 1992. *American National Standard for Disinfection of Water–Storage Facilities.* ANSI/AWWA C652-92. Denver, CO. AWWA.

AWWA (American Water Works Association). 1997. *American National Standard for Disinfection of Water Treatment Plants.* ANSI/AWWA C653-97. Denver, CO. AWWA.

AWWA (American Water Works Association). 1999. *American National Standard for Disinfecting Water Mains.* ANSI/AWWA C651-99. Denver, CO. AWWA.

Bean, S.L. 1995. Dechlorination Alternatives. *Environmental Protection*, 6(5): 25-26.

Best Sulfur Products. 1997. CAPTOR. Product Bulletin. Fresno, CA.

BetzDearbon. 1997. Sodium Bisulfite: Chlorine Removal. Product Bulletin. Trevose, PA.

Bockman, E. 1997. Catalytic Carbon: A New Weapon in the Chloramine Battle. *Water Conditioning and Purification* 39 (7):36-38.

Connell, G.F. 1996. *The Chlorination/Chloramination Handbook.* American Water Works Association. Water Disinfection Series. Denver, CO.

Eckrich, J.A. 1997. Portable Dechlorinator Takes Top Honor. *Opflow* 23(8):4. August. American Water Works Association.

Environmental Compliance Section, East Bay Municipal Utility District. 1998. *Environmental Impacts of Potable Water Discharges.*

Environmental Engineering and Science Section, Washington Suburban Sanitary Commission. 1998. *Final Report, Best Management Practices Plan for Potable Water Discharges.*

Farmer, R.W., and Kovacic, S. L. 1997. Catalytic Activated Carbon Offers Breakthrough for Dialysis Water Treatment. *Dialysis & Transplantation.* November.

General Chemical. 1988. Water Treatment Dechlorination. Product Bulletin. Parsippany, NJ.

GVRD (Greater Vancouver Regional District). 1993. *Environmental Impact Assessment of proposed Secondary Disinfection of Drinking Water: Stage II.* Vancouver, Canada.

GVRD (Greater Vancouver Regional District). 1997. *Agricultural Water Use Guidelines.* Vancouver, Canada.

GVRD (Greater Vancouver Regional District). 1997. *Best Management Practices for Reservoir Cleaning.* Vancouver, Canada.

GVRD (Greater Vancouver Regional District). 1997. *Best Management Practices for Pigging and Flushing Water Mains.* Vancouver, Canada.

GVRD (Greater Vancouver Regional District). 1997. *Construction Water Use Guidelines.* Vancouver, Canada

GVRD (Greater Vancouver Regional District). 1997. *Generic Emergency Response Plan for Chlorinated Water.* Vancouver, Canada

GVRD (Greater Vancouver Regional District). 1997. *Industrial/Commercial Water Use Guidelines.* Vancouver, Canada.

GVRD (Greater Vancouver Regional District). 1997. *Municipal Water Use Guidelines.* Vancouver, Canada

GVRD (Greater Vancouver Regional District). 1997. *Chlorine Monitoring and Dechlorination Techniques Handbook.* Vancouver, Canada

Hach Company (1995). Current Technology of Chlorine Analysis for Water and Wastewater. *Technical Information Series - Booklet No. 17.* Golden, Colorado.

Hardison, R. L. and Hamamoto, M. (1998). An Alternative Dechlorination Process. CWEA Annual Conference, Oakland, CA.

Hardison, R. L. (1999). Best Sulfur Products. Modesto, CA. Personal Communication.

Helz, G. R., and Nweke, A. C. 1995. Incompleteness of Wastewater Dechlorination. *Environmental Science & Technology,* 29 (4): 1018-1022.

Mallinckrodt Baker. 1999. Material Safety Data Sheet – Sodium Sulfite. Phillipsburg, NJ.

Metcalf and Eddy, Inc. 1981. *Wastewater Engineering: Treatment, Disposal and Reuse.* McGraw Hill, New York.

Peterka, G. 1998. Vitamin C - A Promising Dechlorination Reagent. *Opflow.* 24 (12): 1-5. AWWA.

Snoeyink, V.L. and Suidan, M.T. 1975. Dechlorination by activated carbon and other reducing agents. In *Water and Wastewater Disinfection.* Johnson, J.D. (Ed). Ann Arbor Science Publishers Inc. Ann Arbor, MI (pp 339-358).

Southern Ionics Incorporated, (1998). Sodium Bisulfite Solution. Application Bulletin. Matthews, NC.

Spotts, S., and McClure, A. (1995). Catalytic/Adsorptive Carbon Creates a Media "Breakthrough". *Water Conditioning and Purification.*

USEPA (United States Environmental Protection Agency). 1984. *Ambient Water Quality Criteria for Ammonia.* EPA 440/5-85-001. Washington, DC. USEPA.

USEPA (United States Environmental Protection Agency). 1986. *Quality Criteria for Water.* Report number EPA 440/5-86-001. Washington, DC. USEPA.

USEPA (United States Environmental Protection Agency) 1988. *Methods for Aquatic Toxicity Identification Evaluations, Phase I, Toxicity Characterization Procedures.* Report No. USEPA-600/3-88/034, pp 8-27 & Table 8.4.

USEPA (United States Environmental Protection Agency) 1999. Update of Ambient Water Quality Criteria for Ammonia. EPA 822-R-99-014. December.

Vasconcelos, J.J. et.al. 1997. Kinetics of chlorine decay. J. AWWA. 89: 54-65.

WPCF (Water Pollution Control Federation) 1986. *Wastewater Disinfection. Manual of Practice FD-10.* Alexandria, Virginia.

White, G. C. 1999. *Handbook of Chlorination and Alternative Disinfectants.* A Wiley-Interscience Publication, New York.

ABBREVIATIONS

APHA	American Public Health Association
AWWA	American Water Works Association
AwwaRF	Awwa Research Foundation
BAT	best available technology
BC	British Columbia
BCT	best conventional pollutant control technology
BDL	below detection limit
BMPs	best management practices
BPT	best practicable control technology
CCC	chronic criterion
CCME	Canadian Council of Ministers of the Environment
CEPA	Canadian Environmental Protection Act
CERCLA	Comprehensive Environment Response, Compensation and Liability Act
CFR	Code of Federal Regulations
CMC	acute criterion
CWA	Clean Water Act.
CWW	Cincinnati Water Works
DO	dissolved oxygen
DPD	N,N-diethyl-p-phenylenediamine
EBCT	empty bed contact time
EBMUD	East Bay Municipal Utility District
ELS	early life stages
g/L	grams per liter

gpm	gallons per minute
GVRD	Greater Vancouver Regional District
kg	kilograms
lb	pound
MDEQ	Montana Department of Environmental Quality
MGD	million gallons per day
mg	milligram
mg/L	milligrams per liter
MSDS	material safety data sheet
MWD	Metropolitan Water District of Southern California
µg/L	micrograms per liter
NFPA	National Fire Protection Association
NPDES	National Pollution Discharge Elimination System
OSHA	Occupational Safety and Health Administration
ODEQ	Oregon Department of Environmental Quality
ORP	oxidation reduction potential
PAC	Project Advisory Committee
PEL	permissible exposure limit
PEI	Prince Edward Island
POTW	publicly owned treatment works
ppm	parts per million
PVC	polyvinyl chloride
RAM	Redox Alloy Media
RMP	Risk Management Plans
RWQCB	Regional Water Quality Control Boards

SARA	Superfund Amendments and Reauthorization Act
SDWA	Safe Drinking Water Act
STEL	short term exposure limit
THMs	trihalomethanes
TLV	threshold limit value
TMB	tetramethylbenzidine
TMDL	total maximum daily load
TRC	total residual chlorine
TWA	time weighted average
U.S.	United States
USDOT	United States Department of Transportation
USEPA	United States Environmental Protection Agency
USFDA	United States Food and Drug Administration
WAC	Washington Administrative Code
WLA	waste load allocation
WPCF	Water Pollution Control Federation
WQ	water quality
WQS	water quality standards
WSSC	Washington Suburban Sanitation Commission